THE FARM TRACTOR HANDBOOK

[THE "NATION'S FOOD" SERIES No. 1.]

BY

GEORGE SHERWOOD.

LONDON:
ILIFFE & SONS LIMITED, 20, TUDOR ST., E.C.4.

CONTENTS.

INTRODUCTION.

FROM John-o'-Groat's to Land's End and from the Atlantic to the North Sea there is scarcely a square mile of fertile countryside that has not during the past eighteen months echoed to the beat of the agricultural tractor, and there are few farmers in this country who have not had opportunities of studying first hand its capabilities. Perhaps, without paying too much heed to the many plausible arguments used in its favour by zealous agents and armchair food producers, the British farmer has gradually gained a very high regard for the tractor, based on his own level-headed judgment and observation, and, whatever his first impressions were, his verdict to-day is that the tractor has made good.

To state here that its adoption means getting the work done more quickly and at the right time, resulting in larger crops, cheaper production, and more profit, would merely be repeating what the farmer has already found out for himself. The object of the author, therefore, has been to begin where most other writers have left off. Instead of informing the farmer why he should buy a tractor, this book is intended to aid him to choose out of the ever-growing number of different makes, a machine that will best meet his particular requirements, and to explain in simple language the manner of its working, how to maintain it in good order, and how to use it to the best advantage. G.S.

CHAPTER I.

THE INTERNAL COMBUSTION ENGINE.

DURING the first half of the eighteenth century many inventors were inspired by the idea that if the explosive that propels a bullet from a gun could be harnessed and made to turn wheels, it would provide a cheap and almost unlimited form of power. None of them really succeeded until, in 1860, Lenoir, a Frenchman, designed a two-stroke engine for which he found a good many buyers. Sixteen years later, however, his triumph began to wane, for in 1876 Dr. Otto patented the four-stroke engine—known as the Otto cycle engine—which is the parent of the stationary gas and oil engine, the motor car engine, the farm tractor engine, and the aero engine of the present day. The internal combustion engine we are about to discuss, therefore, has some of the characteristics of the gun—that is, a short wide barrel (the cylinder) is provided, one end of which is closed, and the other, being open, is plugged by a closely-fitting but movable wad or bullet (the piston), and by an arrangement of valves the explosive is introduced into the closed end of the cylinder and is fired so as to propel the piston forward. Instead of being shot out of the cylinder, the piston, as it is propelled forwards, expends the force of the explosion in turning a crankshaft (by means of a connecting rod) to which is attached a flywheel, and from this crankshaft and flywheel the power is easily carried to where it is wanted.

Of course many devices have to be introduced for providing the explosive vapour, for firing it at exactly the right moment, for getting rid of the exploded fumes, for continuing the explosions regularly, for lubrication, for preventing undue heating of the parts affected by the explosions, and for the conveyance of the power to where it is wanted.

It is these devices which make the engine of a motor car or tractor look complicated, but as a matter of fact it will be found comparatively simple if the principles on which it works are thoroughly understood. The only fundamental difference between a gas engine and an oil engine is that the gas engine takes its explosive directly in the form of gas, while the oil engine has to vaporise its oil to make its explosive. For the benefit of those unacquainted or only slightly acquainted with internal combustion engines, some of their parts will be briefly explained.

How the Engine Works.

The engine, it will now be understood, is that part of the tractor where the power is actually generated, and, though it does not constitute the whole of the mechanism of the complete machine, it is naturally the most important part. Briefly, the power is obtained by exploding, or burning, a mixture of liquid fuel vapour and air in one or more closed chambers, termed cylinders, the explosions acting upon the pistons therein, which are consequently pressed outwards, causing a shaft to rotate. This rotation is communicated at will to the driving wheels, turning them at various speeds.

The engine requires for its operation certain accessories, each of which will be dealt with in turn. The first of these is the carburetter, or mixer, which supplies it with fuel vapour; the second, the electric magneto machine which fires or ignites the gas or vapour; the third, the water-cooling system, to keep the temperature down to a working heat; and the fourth, the lubricating system.

In the description of the engine each of these accessories will be taken for granted, and it must, therefore, for the time being, be assumed by the reader that means are provided for supplying gas to the engine, igniting it, etc. The reader's attention is now directed to fig. 1, which shows the main mechanical features of the engine in their very simplest form. The upper part of this illustration is a sectional view; that is to say, it represents the engine cylinder cut in half vertically. The lower part of the illustration shows a sketch of the connecting rod, crank, crankshaft, and flywheel.

The cylinder consists of a tube, the upper end of which is closed and the lower open, but is not in actual fact of the simple shape shown in fig. 1. Its inner bore or surface is machined very accurately to a certain diameter, which is termed the "bore" of the engine. Sliding up and down in the cylinder is a piston, which closely fits the walls of the cylinder. Round the upper part of the piston are one or more split rings, termed piston rings, which spring outwards against the cylinder walls and prevent any leakage of gas or air past the piston. The piston has attached to it a connecting rod, the lower or "big end" of which takes a bearing on the crank pin. The crank pin is carried by cranks on a shaft, which is termed the crankshaft. If the crankshaft is rotated by hand in the direction of the arrow, the connecting rod will first pull the piston down in the cylinder and then push it up again. The rotation of the crankshaft, therefore, reciprocates the piston up and down in the cylinder. Similarly, if the parts were in the position shown in fig. 1, and pressure were applied to the top of the piston, the connecting rod would push the crank downwards, and the flywheel and the crankshaft would be rotated in the direction of the arrow.

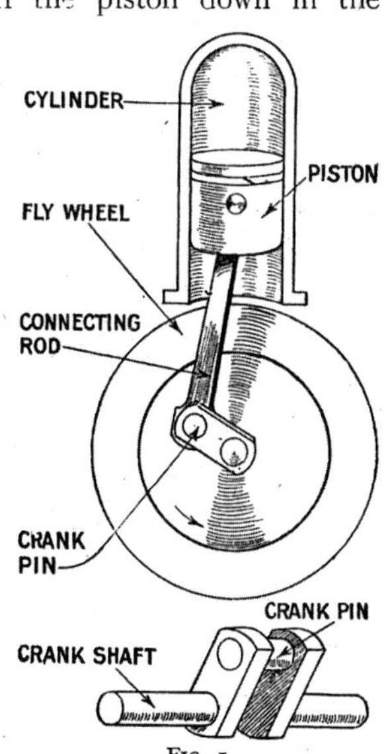

Fig. 1.
Diagrammatic sketch of cylinder, piston, connecting rod, and crank.

If now the parts are in the position shown, and some gas were in the top part of the cylinder, which is called the combustion chamber, and if the gas were exploded, the explosion pressure would be exerted on top of the piston, which would cause the crankshaft to rotate in the direction of the arrow. In this way the explosion obtained from the gas causes the crankshaft to turn. The cylinder, after one impulse of this kind, must be refilled with explosive gas, the piston, after the burnt gas has been removed or exhausted, must be brought into the position shown, and the parts generally prepared for a fresh explosion. The manner in which this is performed will now be described.

The Cycle of Operations.

The filling of the cylinder with fresh gas, and the exhausting of the burnt or dead gas, is automatically effected by means of valves, which are shown in detail in figs. 2 and 3 and in operation in figs. 4 to 7.

Returning to fig. 1, it was stated that the cylinder was not of the simple regular shape illustrated therein, as is

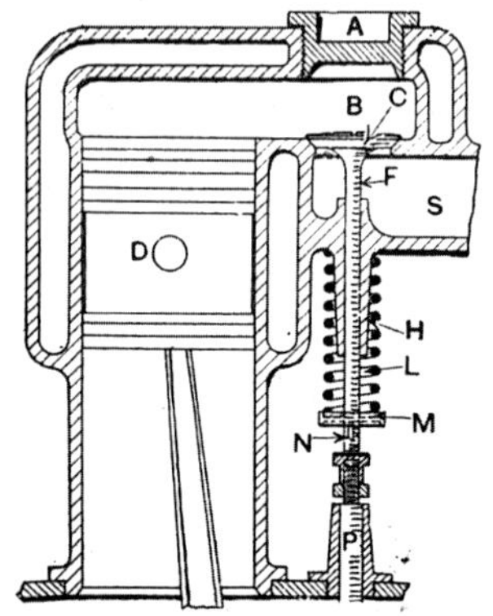

FIG. 2.
SECTION OF CYLINDER PROVIDED WITH STANDARD TYPE POPPET VALVE.

A. Valve cap.
B. Valve pocket.
C. Valve head.
D. Piston.
F. Valve stem.
H. Valve guide.
L. Valve spring.
M. Valve washer.
N. Valve cotter.
P. Valve tappet.
S. Gas inlet from carburetter or outlet from cylinder.

clear from fig. 2, in which is shown a single cylinder with the valve in place, whilst in fig. 3 it is shown removed. The cylinder is formed with a pocket B at one side, and on the underside of this pocket there is a hole E (fig. 3), the upper

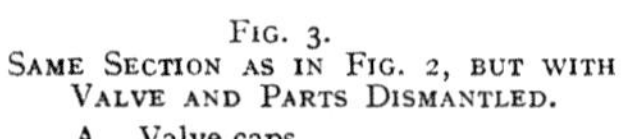

FIG. 3.
SAME SECTION AS IN FIG. 2, BUT WITH VALVE AND PARTS DISMANTLED.

A. Valve caps.
C. Valve head.
E. Valve hole.
F. Valve stem.
H. Valve guide.
L. Valve spring.
M. Valve washer.
N. Valve cotter.
O. Tappet adjustment.
P. Valve tappet
R. Valve seat.

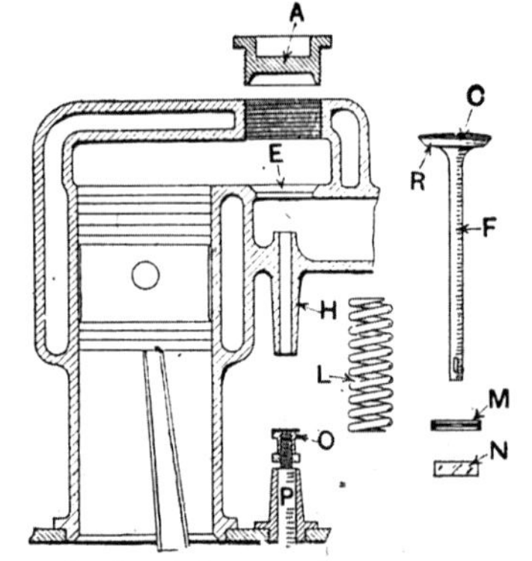

surface of which is formed conical. The valve pocket B and the interior of the cylinder therefore communicate with the passage S through the hole E E. Resting in the hole E is a disc C, coned on the underside R and attached to a stem F, this stem being acted upon by a spring L, so as to hold the valve disc or head C down into the hole, and cut off communication between the inside of the cylinder and the passage S. There are two of these valves—one on each side of the cylinder, or one behind the other, or one above the other, or both in the cylinder head, as in fig. 8. One is used for the admission of fresh gas, and is termed the inlet valve, whilst the other, which is termed the exhaust valve, is employed for releasing the burnt gas.

We will assume that the cylinder shown in fig. 2 is to be filled with gas, and that the piston D is about to descend. The inlet valve is, therefore, now opened by a stem P, termed a valve tappet, rising and pushing up the valve stem, thus lifting the valve head C from its seating in the hole E. (This is what has occured in fig. 4.) As the piston descends, it can now suck in fresh gas along the passage S (fig. 2). At the right time the valve tappet P drops automatically, and the valve is then enabled to be closed by the spring: whilst later, at the end of the working stroke, another tappet rises and lifts the exhaust valve. Each valve is lifted by a cam, which consists briefly of a projection on a rotating shaft, as can be seen by M and N in figs. 4 to 7. Figs 4 to 7 represent an actual engine cut vertically through the centre of one of the cylinders so as to expose its interior mechanism. In this case the two valves are on opposite sides of the cylinder, the inlet valve F being on the left and the exhaust valve G on the right. It will be seen that the upper end of the cylinder A is closed, while the connecting rod J passes through the other end, which is open, on its way from the piston C to the crank pin on the crankshaft K. In the closed end of the cylinder are the two valves F and G. F is the inlet valve, and the explosive gas is led to it through a pipe and passage I. G is the exhaust valve, and the waste gases pass out through it and the passage H to the exhaust box or silencer (not shown), and thence into the open air.

The engine has a cycle or series of four operations: (1.) Induction or inlet. (2.) Compression. (3.) Firing or combustion. (4.) Exhaust. All are performed in two revolutions of the crankshaft, corresponding to two outward or downward strokes, and two return or upward strokes of the piston.

(1.) *Induction.*—Suppose the parts to be in the position shown in fig. 4, and the crankshaft to be rotated, the crank will descend and draw down the piston C by means of the

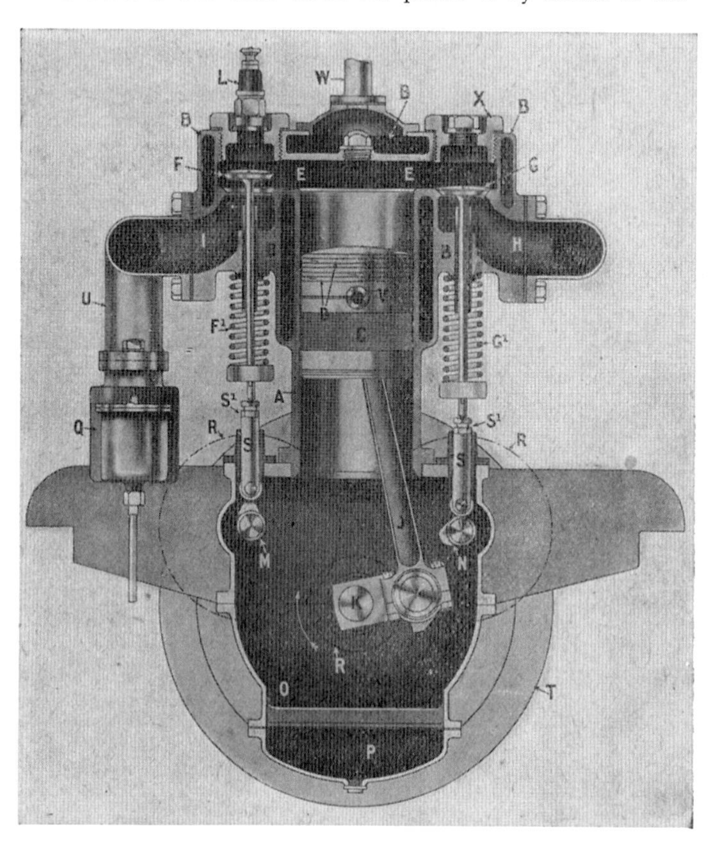

FIG. 4. THE CYCLE OF ENGINE OPERATIONS. (1) THE INDUCTION OR INLET STROKE. INLET VALVE F OPEN PISTON D DESCENDING.

A. Cylinder. C. Piston. E. Combustion chamber.
B. Water jacket. D. Piston rings. F. Inlet valve (open).

connecting rod J. As the piston descends, it tends to create a vacuum or suction in the space E, which is called the combustion chamber. The inlet valve F is at the same time pushed open, being raised by means of the cam M, and the combustible gas is drawn into the space E past the inlet valve F. As the piston approaches the end of its

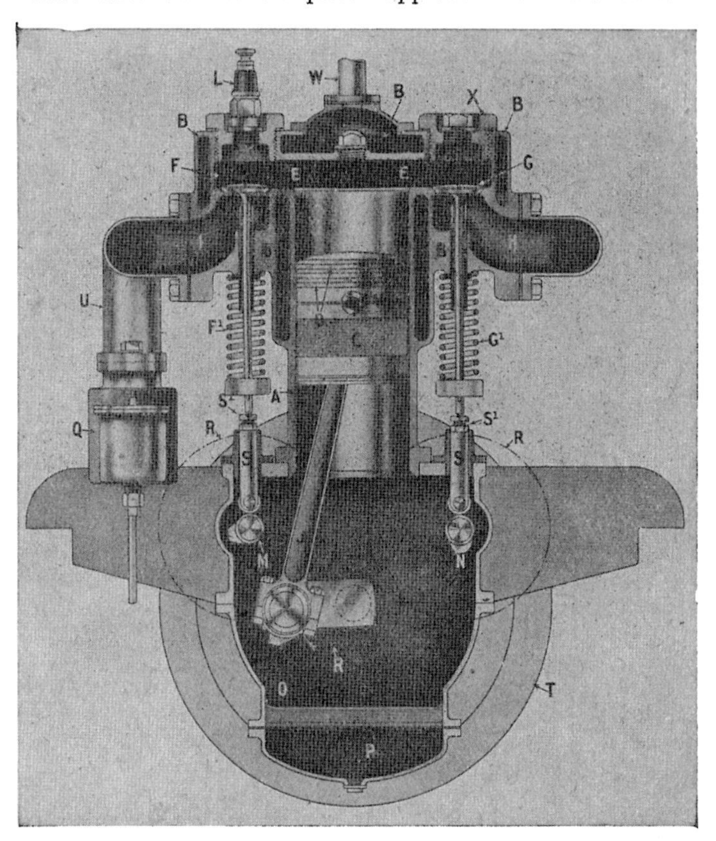

FIG. 5. THE CYCLE OF ENGINE OPERATIONS. (2) THE COMPRESSION STROKE. INLET VALVE F AND EXHAUST VALVE G CLOSED, PISTON D ASCENDING.

F^1. Inlet valve spring. G^1. Exhaust valve spring. I. Inlet port.
G. Exhaust valve. H. Exhaust outlet.

outward stroke, the valve F is closed by its spring F1, and the gas is imprisoned in the combustion chamber E. But it is no use firing it yet, because it could not drive the piston any further if this were done, so we must cause the piston to return to its other extreme position. We must not fire the gas before the piston reaches the top of the cylinder,

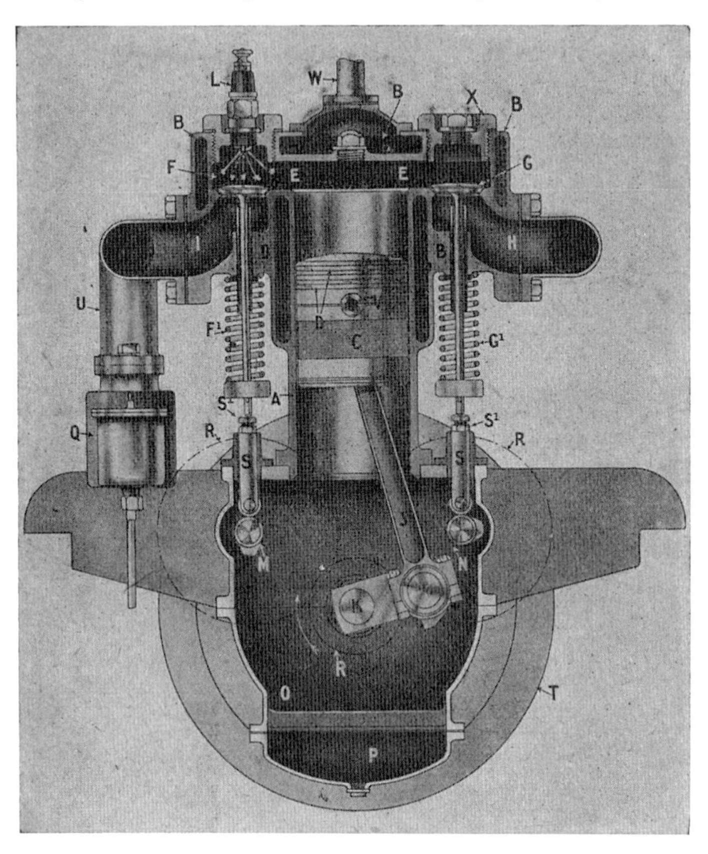

FIG. 6. THE CYCLE OF ENGINE OPERATIONS. (3) THE FIRING STROKE. INLET VALVE F AND EXHAUST VALVE G CLOSED, PISTON D DESCENDING.

J.	Connecting rod.	N.	Exhaust valve cam.
K.	Crankshaft.	O.	Crank chamber.
L.	Sparking plug.	P.	Oil sump.
M.	Inlet valve cam.	Q.	Carburetter or mixer.

or we shall make the piston return too soon and drive the crankshaft round in the reverse direction.

(2.) *Compression.*—As the rotation of the crankshaft is continued, the piston rises and compresses the gas in the combustion chamber (see fig. 5).

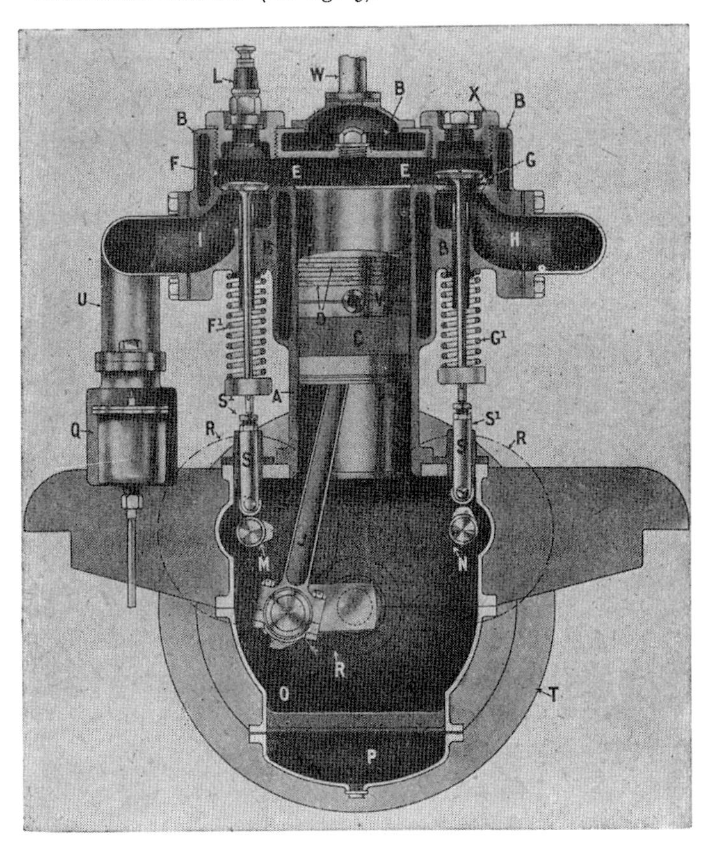

FIG. 7. THE CYCLE OF ENGINE OPERATIONS. (4) THE EXHAUST STROKE. EXHAUST VALVE G OPEN, PISTON D ASCENDING.

R. Timing pinions.
S. Valve tappets.
S1. Valve tappet adjustments.
T. Flywheel.
U. Inlet pipe.
V. Gudgeon pin.
W. Water outlet.
X. Valve cap.

(3.) *Firing.*—When the piston has got back to the top of its stroke, the compressed gas is fired by an electric spark at the points of the sparking plug L. The consequent explosion or combustion exerts pressure on top of the piston, which drives it along the cylinder with great force. When the piston has nearly reached the bottom of this stroke, the exhaust valve G is opened mechanically by means of the cam N, and the waste gases begin to rush past it and out through the passage H to the silencer (fig. 6).

(4.) *Exhaust.*—As the piston rises for the second time, it sweeps out nearly all the remaining waste gases through the open exhaust valve G (fig. 7). By the time the piston

THE AVERY TRACTOR. This Machine is made in Three Sizes—A, B, and C.

	A.
Engine.	Four cylinders, 3 × 4in. (76 × 102 mm.)
Power.	5 h.p. (1,000 lb., assuming 200 lb. per h.p.), 10 h.p. at belt pulley.
Fuel.	Petrol.
Weight.	20 cwt. 0 qr. 10 lb.
Capacity.	Two-furrow plough.
Wheels.	Four. Back, 3ft. 2in. × 5in. face; front, 2ft. 4in. × $\frac{3}{4}$in. face; square outer rims.
Overall Dimensions.	Length 12ft. 1in., wdith 2ft. 11in.
	B.
Engine.	Two cylinders, $5\frac{1}{2}$ × 6in. (140 × 152 mm.)
Power.	8 h.p. (1,600 lb., assuming 200 lb. per h.p.), 16 h.p. at belt pulley.
Fuel.	Paraffin.
Weight.	43 cwt. 3 qr.
Capacity.	Three-furrow plough.
Wheels.	Four. Back, 4ft. 2in. × 12in. face; front, 2ft. × 6in. face.
Overall Dimensions.	Length 10ft. 10in., width 4ft. 8in., height 4ft. 5in.
	C.
Engine.	Two cylinders, $6\frac{1}{2}$ × 7in. (165 × 178 mm.)
Power.	12 h.p. (2,400 lb., assuming 200 lb. per h.p.), 25 h.p. at belt pulley.
Fuel.	Paraffin.
Weight.	67 cwt.
Capacity.	Four-furrow plough.
Wheels.	Four. Back, 4ft. 8in. × 1ft. 8in. face; front, 2ft. 6in. × 8in. face.
Overall Dimensions.	Length 13ft. 8in., width 6ft. 8in., height 8ft. 9in.

has reached the end of this stroke, the cam N has moved away, allowing the exhaust valve to return to its seat ready for the next suction stroke.

THE BATES STEEL MULE AGRICULTURAL TRACTOR.
THE SPECIFICATION OF THIS MACHINE WILL BE FOUND ON PAGE 22.

This series or cycle of operations is repeated over and over again, recurring anything from one hundred to five hundred times a minute in each cylinder, according to the type of engine and the amount of gas allowed past the throttle, which is controlled by hand.

As each valve only lifts once in every two revolutions, the camshafts carrying the cams M N must only revolve at half the speed of the crankshaft. This reduction of speed

THE BATES STEEL MULE AGRICULTURAL TRACTOR.

ENGINE.	Four cylinders, 4 × 6in. (102 × 152 mm.)	FUEL.	Paraffin or petrol.
POWER.	16 h.p. (3,200 lb., assuming 200 lb. pull per h.p.), 30 h.p. at belt pulley.	CAPACITY.	Three or four-furrow plough, self-binders, mowers, etc., or full-sized threshing machines.

WHEELS. Tread width adjustable from 3ft. 9in. to 6ft. 10in; two front steering wheels and central rear drive by means of chain tracks.

is obtained by fitting the camshafts with toothed wheels R R (outlined in figs. 4 to 7) meshing with a similar wheel on the crankshaft with only half the number of teeth. The magneto and water pump are usually driven off this train of wheels, which is termed the timing gear. The setting of these wheels determines the moment at which the valves come into operation in relation to the position of the pistons, also the time at which the spark takes place at the sparking plug. It will be understood, therefore, that the correct setting is very important. Certain teeth are usually marked by the makers, so that the correct setting can be easily found if for any reason the timing gear is ever dismantled.

Types of Engine.

Figs. 1 to 7 represent the ordinary poppet valve engine. This type is almost exclusively used on both British and American tractors; the only variation from the illustrations (which show vertical engines) being that many tractors are

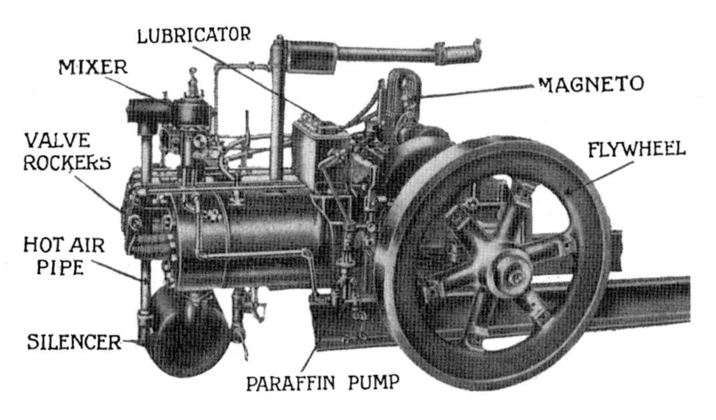

Fig. 8.
A typical twin-cylinder horizontal engine of the unopposed cylinder model.

equipped with horizontal engines, as shown in fig. 8. Sometimes the cylinders of these horizontal engines are opposed —that is, one cylinder is placed each side of the crankshaft. This arrangement of a horizontal engine gives greater freedom from vibration and permits of a higher engine speed.

Besides the poppet valve engine, there are sleeve valve engines, rotary valve engines, two-stroke engines, and Diesel engines. All of these may sooner or later make their appearance on the land. When this does occur, revised editions of this book will deal with them. In the meantime, those interested will find them very fully described and illustrated in "The Autocar Handbook," a sister publication.

Cranks and Bearings.

The crankshaft is mounted in bearings in the crank chamber, which is usually of aluminium or cast iron, and comprises a mere shell adapted to retain lubricating oil, and is strong enough to carry the crankshaft, and have the cylinders bolted to it. The number of bearings for a crankshaft may vary. In the case of a single-cylinder engine there are, of course, two bearings, whilst two or three may be employed for a two-cylinder engine. For a four-cylinder engine there are two, three, or five. The more bearings the better, as a rule. Some of the best two-cylinder engines, however, only have two, but less than three in a four-cylinder engine is a weakness.

Pistons.

Pistons are usually made of cast iron, and each is fitted with a number of rings, which possess a certain amount of spring, and expand against the cylinder wall so as to form a gastight joint, in order to prevent a leakage of gas past the sides of the piston.

The Silencer.

If the exhaust gases pass straight from the engine into the open air, considerable noise is created. To prevent this, the gases are led through a silencer, which comprises a box, usually cylindrical in shape, formed with a number of compartments, in the walls of which are drilled small holes through which the exhaust gas issues or percolates, so to speak, in a number of small streams.

The exhaust gases should be free from smoke. A light blue smoke either indicates over-lubrication or imperfectly vaporised paraffin. The latter will occur if the paraffin be turned on too full, or turned on before the vaporiser is quite warm. Blue smoke may also indicate that the water drip is being used too liberally, causing the mixture to be cooled down more than is necessary. Black pungent smoke, accompanied by a knocking noise in the engine, indicates pre-ignition—a condition which can be obviated by a freer use of the water drip.

BULLOCK CREEPING GRIP TRACTOR.

ENGINE.	Four cylinders (in pairs), $4\frac{1}{2} \times 6\frac{3}{4}$in. (114 × 171 mm.)	GEARS.	One forward, one reverse.
POWER.	$12\frac{1}{2}$ h.p. (2,500 lb., assuming 200 lb. pull per h.p.), 30 h.p. at belt pulley.	CAPACITY.	Four furrows under practically all conditions.
FUEL.	Petrol, paraffin, alcohol, or distillate.	TRACKS.	Two. Width of track 12in., length 7ft.; effective tread on ground over 8 sq. ft.
WEIGHT.	65 cwt., packed for export 80 cwt.	OVERALL DIMENSIONS.	Length 9ft., width 6ft. 9in., height 6ft. 4in.

Sometimes explosions take place in the silencer. These explosions indicate irregular running of the engine, which is often caused by the mixture being too weak. Consequently it will not explode with the spark in the cylinder, and finds its way into the silencer, where the red hot gases from the next explosion in the cylinder enter and fire it with a sharp report. This trouble may be overcome by careful adjustment of either the air valve spring or the fuel valve. If this does not remedy matters there is probably a defect in the ignition—most likely a sparking plug—which results in gas, which is misfired in a cylinder, being subsequently exploded in the silencer in the same manner as described above.

CHAPTER II.

CARBURATION, LUBRICATION, AND COOLING.

Fuel.

ALL grades of petrol, benzole, paraffin, and many other hydro-carbon oils are suitable fuel for the internal combustion engine. The lighter oils, such as petrol and benzole, will evaporate without the aid of artificial heat, and will start an engine from cold, but the heavier, less volatile oils, like paraffin, need heat to convert them into vapour.

As paraffin is the cheapest and easiest to obtain of all liquid fuels, it is almost exclusively used on farm tractors, and performs very satisfactorily, its only disadvantage being that it will not start the engine from cold, which makes it necessary to use a little petrol for starting.

Coal gas is almost an ideal fuel, and at its present price it would be a very attractive substitute but for the fact that it is so cumbersome to handle. It must either be stored in a large canvas container at atmospheric pressure, or compressed into strong heavy steel cylinders. About 250 cubic feet are equal to one gallon of paraffin, and a good sized farm waggon would carry a container large enough to hold 1,000 cubic feet, while a cylinder 49in. × 7in., weighing 108 lb., would hold about 130 cubic feet at 1,800 lb. pressure per square inch. Gas in a container would be quite practicable for stationary work, and even desirable where a farm was situated near a gasworks. Certain small alterations have to be made to the carburetter, which, however, need not interfere with the paraffin arrangements. Where a farm was supplied with gas from the main it would be foolish to use anything else for stationary work, for at 3s. per 1,000ft. it is equivalent to paraffin at 9d. per gallon.

Carburation.

It has already been explained that the source of energy developed in the cylinders is a mixture of fuel vapour and air. To obtain the maximum of power the mixture must contain a given quantity of each (about one part fuel vapour and fifteen to seventeen parts air), and it is the function of the carburetter to provide this mixture and supply it in the correct proportions.

As paraffin will not evaporate at ordinary atmospheric temperatures, before it can be converted into explosive gas it must be heated. The handiest form of heat to utilise is obviously the hot waste gases passing through the exhaust pipe, and this is done with the aid of a vaporiser. However, until the engine is running, the necessary heat is non-existent; hence the necessity of using petrol for starting.

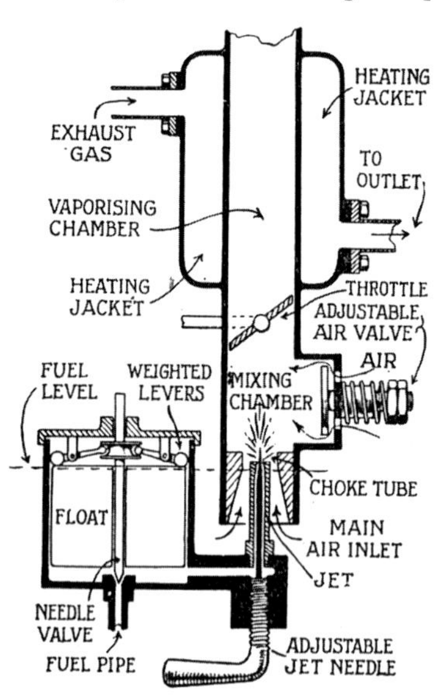

FIG. 1.
Sectional view showing general arrangement of a standard type carburetter and vaporiser in which the heat of the exhaust gas is utilised to assist the vaporisation of the fuel.

Carburetters or Mixers.

Though different makes of carburetter vary greatly in detail, all in general use work on the same principle. As space forbids a description of specified makes, a simple drawing is shown in fig. 1 which embodies the essential features of them all. It will be seen that a spraying jet is placed in a mixing chamber, which is provided with ports, or openings, for the admission of air, and that the jet is in communication with a chamber containing a float (the float chamber). Either petrol or paraffin is admitted at will to the bottom of the float chamber by means of pipes leading from the fuel tanks. The tanks are placed on a higher level than the carburetter, so that the fuel flows by its own weight. When the fuel reaches a certain height in the float chamber, the top of the float comes in contact with two weighted levers, which are pivoted to the under-side of the lid. The counter-weights on these levers are normally enough to keep a needle valve raised, but as the fuel flows into the float chamber and the float moves upwards, it comes into contact with the bob-weights on these levers, and by its buoyancy counteracts the effect of the weights, so that they are no longer able to keep the needle valve raised. The latter accordingly falls of its own weight. Then as the fuel is used up by the engine the float falls, once more raising the needle, and by this means a constant level is always maintained.

Extending horizontally from the float chamber is a passage, ending in a vertical jet tube, and the height of the tube is just a shade above the level at which the fuel is maintained by the float. Consequently, fuel is always standing in the jet ready to be used by the engine. Air is admitted to the mixing chamber through a tube with a restricted opening which surrounds the jet. This tube is termed the choke tube. When the engine is turned, air is sucked in through the choke tube with great velocity, and this causes fuel to issue from the jet in the form of fine spray. The spray in the case of petrol quickly turns to vapour, and mingles with the air. In the case of paraffin, it is converted into vapour in the hot vaporising chamber through which it subsequently passes.

Correct Mixture.

As a correct explosive mixture is most important, it will be apparent to the reader that means must be provided for admitting fuel and air in the right proportions under varying conditions of load and engine speed. In some cases the object is attained by placing a needle valve in the jet orifice, which may be turned at will to vary the size of the orifice. In others, an adjustable spring-controlled air valve is provided, which automatically admits more air as the engine goes faster and the suction increases. Sometimes both an air valve and jet needle are supplied. By careful adjustment of one or both, the correct setting for giving a good mixture can quickly be found.

A valve (the throttle valve) is placed between the carburetter and the engine to regulate the supply of gas to the cylinders, and this is controlled by a lever placed conveniently near the driver. The type of carburetter described, in various forms, is almost universally used on motor cars of all descriptions, and is extremely reliable.

The only troubles that might be anticipated are:

(1.) Stoppage of the fuel through dirt or water getting in the jet or supply pipes.

(2.) Flooding caused by the float needle sticking or not being a good fit in its seating.

(3.) Flooding caused through the float being punctured and getting partly filled with fuel.

THE 18 H.P. "CATERPILLAR" TRACTOR.

ENGINE.	Four cylinders, $4\frac{1}{2} \times 6$in. (114 × 152 mm.)	WEIGHT.	54 cwt.
POWER.	12-16 h.p. (assuming 200 lb. pull per h.p.), 18 h.p. at belt pulley.	GEARS.	Three forward and one reverse; speeds in m.p.h. at 750 r.p.m., 1.5, 2.2 and 3.6.
		CAPACITY.	Three furrows.
FUEL.	Paraffin or petrol.	TRACK.	11in. wide, track each side.

OVERALL DIMENSIONS. Length 9ft. 6in., width 4ft. 3in., height without canopy 4ft. 8in., length of ground contact 5ft. 4in.

The chances of a stoppage in the pipes or jet are reduced to a minimum by a fine gauze strainer, which is placed somewhere between the fuel tank and the carburetter. This strainer should be located so as to be easily removable, and should be occasionally cleaned.

If, however, the pipes or jet do get stopped, the pipes can be easily disconnected and blown through, and the jet is always removable, but often requires a special key supplied with the carburetter; this key should always be kept handy.

Flooding is nearly always caused by a little grit getting on the needle valve seating, and a few turns of the needle with the thumb and finger will usually remedy matters. If,

THE 45 H.P. "CATERPILLAR" TRACTOR.

ENGINE.	Four cylinders, 6×7in. (152 × 178 mm.)	GEARS.	Two forward and two reverse; speeds in m.p.h. at 600 r.p.m., $2\frac{1}{8}$ and $3\frac{1}{2}$ and $2\frac{1}{8}$ and $1\frac{1}{2}$.
POWER.	25-30 h.p. (assuming 200 lb. pull per h.p.), 45 h.p. at belt pulley.	CAPACITY.	Five to six furrows.
FUEL.	Paraffin or petrol.		
WEIGHT.	116 cwt.	TRACK.	13in. wide, track each side.

OVERALL DIMENSIONS. Length 12ft. 9in., width 6ft. $2\frac{1}{2}$in., height over canopy 5ft. 10in., length of ground contact 6ft. 8in.

however, it is caused by the seating being worn or faulty, the valve should be ground in by smearing the point of the needle with a little very smooth crocus powder, or knife powder and oil, and turning it in its seating with the thumb and finger, or screwdriver if a slit is provided in the top of the needle.

If the float leaks, the presence of fuel inside it can be detected by shaking it, and it will usually be discovered that the hole through which it has made its way is very tiny. To get the liquid out quickly, it is advisable to puncture another small hole on the opposite side of the float (with a sewing needle) and blow out the liquid, and afterwards close up the holes with just a touch of solder. As a float is usually made of very thin sheet brass, this is not difficult. Another way is to immerse the float in a basin of boiling water with the puncture downwards. As the air inside expands with the heat, the liquid will be forced out by the pressure.

A slightly modified type of carburetter, often called a mixer, which had its origin with the stationary oil engine, is to be found on some of the heavier single and two-cylinder tractors. Here the float chamber is replaced by a small tank, to which the paraffin is pumped mechanically. The pump is set to deliver more paraffin than is required, and when this reaches the correct level the surplus runs into a return pipe, which takes it back to the main tank. The tank in this case is on a lower level than the carburetter.

Enough petrol for starting is placed in a small auxiliary tank and admitted to the jet by a two-way tap.

Vaporisation.

A vaporiser is shown in the upper part of fig. 1. It consists of a chamber (through which the hot exhaust gas passes) surrounding the induction pipe, which, after the engine has been running a little while on petrol, becomes sufficiently heated to evaporate the finely-sprayed paraffin that is passing through.

Sometimes the air is heated before it reaches the carburetter by drawing it through a chamber surrounding the exhaust pipe, and occasionally the paraffin is heated by coiling the feed pipe around the exhaust pipe.

Some tractors work very well indeed on paraffin, and some rather indifferently. This not only proves that paraffin is a very good fuel if the machine is properly adapted to its use, but also proves that some makers have a good deal to learn about vaporisation, compression, and cooling, all of which are important matters in a paraffin engine. It would seem at first sight that the only necessity for converting paraffin into a perfect fuel was sufficient heat in the vaporiser, but that is not so.

Perfect vaporisation can only be had by sacrificing other advantages, and this makes it necessary to strike a compromise. First of all, both paraffin vapour and air become considerably expanded or rarefied by the heat of the vaporiser; therefore, a cylinder full of such expanded gas does not represent as much power when exploded as would a cylinder full of petrol gas, for instance, which can be admitted in a much cooler state. Then again, as the gas is already very hot before being compressed, and compression heats it still

THE CLYDESDALE AGRICULTURAL TRACTOR.

ENGINE.	Four cylinders, $4\frac{1}{4} \times 5\frac{1}{2}$in. (108 × 140 mm.)	WHEELS.	Four. Rear, 5ft. diameter × 10in. face; front, 3ft. 4in. diameter × 4in. face.
POWER.	12 h.p. (2,400 lb., assuming 200 lb. pull per h.p.), 25 b.h.p. at belt pulley.	CAPACITY.	Three-furrow plough, belt work, and hauling.
FUEL.	Petrol.	OVERALL DIMENSIONS.	Length 12ft., width 6ft., height 5ft. 6in.
WEIGHT.	$46\frac{1}{2}$ cwt.		
GEARS.	Two forward, one reverse.		

more, the temperature rises to such an extent that if compression is carried beyond a certain limit spontaneous combustion takes place—that is, the gas is exploded by its own heat. This is known as pre-ignition, and is quite a common occurrence in a paraffin tractor. Its symptoms are a pounding, knocking noise within the cylinders when the machine encounters heavy pulling. To reduce this tendency, manufacturers are obliged to keep the compression below what is the most efficient point as regards power, and finally to overcome it a little water vapour is mixed with the explosive charge of many engines. This is conveyed to the induction pipe by what is usually called a water drip.

The Water Drip.

As compression causes heat, so expansion reduces the temperature; therefore, as the water rapidly expands into vapour, it has the effect of drawing off some of the heat from the explosive charge before it is compressed.

Some water drips consist simply of a small pipe, fitted with a tap, leading from the cooling system to the induction pipe. As it is not wise to give the engine any more water than is necessary to stop the knocking, the tap should be carefully regulated, and turned off completely when running idly or when doing light work.

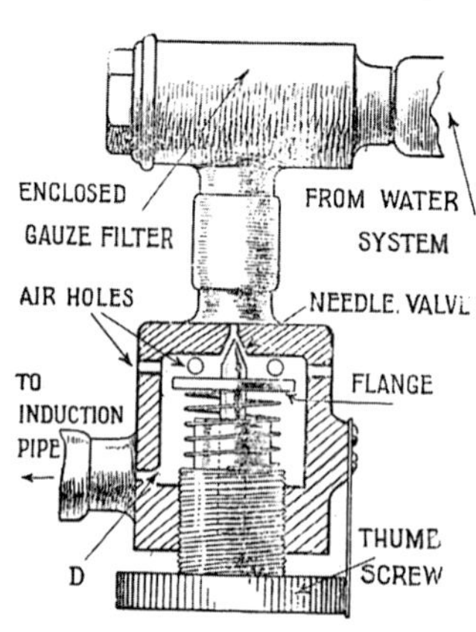

Fig. 2.
Section of water valve. This device allows a small quantity of cold water to be drawn into the cylinder at each suction stroke of the piston. The steam generated in the cylinder reduces the force of the explosion when desired, and so obviates thumping and knocking of an overheated engine.

With the object in view of making the water drip automatically, some makers fit a spring-controlled valve to regulate the supply, a valve of this type being shown in fig. 2. The lower part of the drawing representing the small valve chamber is a sectional view, and shows that the needle valve is held against its seating by a light spring, adjustable by a thumbscrew, and acting against a flange almost equal in diameter to the inside of the chamber. As the engine takes in gas, the suction through the port D draws down the flange, removes the valve from its seating and allows water to enter; air entering through the holes above prevents the water from being drawn into the cylinders with a rush.

The Air Cleanser.

Land cultivation in dry weather is a dusty job, and to guard against dust and other solid matter being drawn into the engine with the air supply, an air washer, scrubber, or filter—of which there are various kinds—is to be found on most tractors. In some cases the air is drawn through a box stuffed with sheep's wool; in others it is drawn down a long chimney-like tube with its open end projecting well above the tractor, where the air is usually comparatively clear. But perhaps the best method of all is that where the air is drawn through a tank of water. This type of

THE CRAWLEY AGRIMOTOR.

ENGINE.	Four cylinders, $4\frac{1}{8} \times 5\frac{1}{2}$in. (105 × 140 mm.)	CAPACITY.	Three furrows, cultivating, binding, mowing, and other field work.
POWER.	30 h.p.		
FUEL.	Paraffin or petrol.	WHEELS.	Three. Two driving wheels 4ft. diameter × 8in. face, one back steering wheel.
WEIGHT.	40 cwt.		
GEARS.	Two forward, one reverse.		

OVERALL DIMENSIONS. Length 17ft. 8in., width 5ft. 6in., height 6ft.

cleanser has the additional advantage that it charges the air with water vapour and makes a water drip unnecessary.

The Governor.

Most tractor engines are provided with a governor to prevent unduly high speeds or "racing," and automatically to regulate the engine to any desired speed, regardless of variations of load. This is a great convenience when doing belt work, as it means that the throttle does not require constant attention when performing such operations as threshing, grinding, sawing, etc., which demand a constant speed with a rapidly varying degree of power.

The governor may be attached to any convenient shaft, and is sometimes for convenience enclosed in the timing wheel case and sometimes attached to the magneto or pump spindle. It usually takes the form of a pair of weighted levers attached to the shaft. The weights are held together by springs, and the levers on which they are mounted are connected to a sliding collar on the shaft. As the engine speed increases the weights fly outwards, causing the throttle to be partly closed through the medium of the sliding collar and levers. When the engine slows down below the desired speed, consequent upon the heavier load being put upon the belt, the weights fall closer together and cause the throttle again to be opened.

Another form of governor is that of the hydraulic type. It consists of either a small piston, or a diaphragm which is acted upon by the cooling water. As the water is circulated more quickly consequent upon increasing engine speed, the piston or diaphragm is pressed outwards, in turn acting upon the throttle through suitable levers.

Obviously a governor of the latter type can only be employed where the water is circulated by a pump, and would not work in conjunction with natural or thermo-syphon circulation.

Lubrication.

If the average British farmer can be said to have a pet aversion, it is to the use of lubricating oil. He spends hundreds of thousands of pounds a year that might otherwise be saved, and countless implement makers have grown rich because he is so sparing with the oilcan. No economy will be recommended here in the use of oil, for a good oil liberally used is an economy in itself, which will reduce depreciation and save many pounds on the repair bill in the course of a year or two. This applies particularly to a tractor engine, for the pistons and various shafts, working at high speeds in such a temperature as prevails, would not work a single minute without oil. The owner of a tractor is therefore urged always to use a good internal combustion engine oil recommended for paraffin engines by a reputable firm.

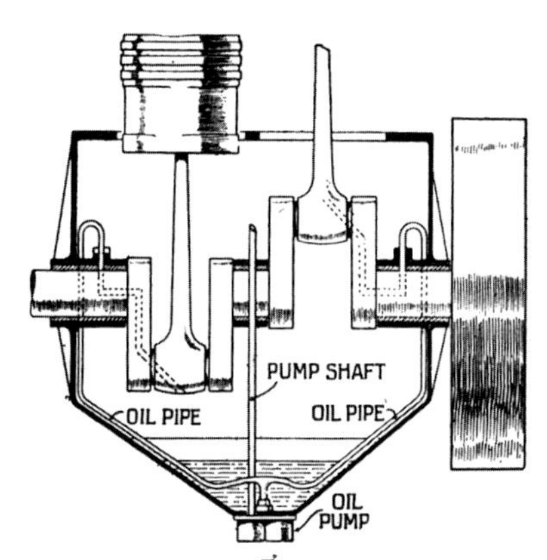

FIG. 3.
Forced lubrication system of a two-cylinder engine. The oil is forced up the oil pipes shown and along the oil ways drilled in the crankshaft to the main and big end bearings, whence it falls out into the crank chamber, and is again pumped round.

There are several systems of engine lubrication in vogue, and, fortunately for the peace of mind of the operator, they are all automatic in action, the only responsibility resting on the driver being that of providing the reservoir with the right oil in the correct quantities, keeping an eye on the sight feed — where one is fitted—to see that the pump is working properly, keeping the oil strainer clean, draining off the old oil at intervals and supplying the reservoir with fresh oil. Beyond this, it can be taken for granted that on a tractor of any repute the oiling system will give little or no trouble.

In most systems a quantity of oil is held in a reservoir, formed by the bottom half of the crank chamber, from which it is pumped by a small mechanical pump to the various bearings. One well-tried and efficient method is shown in fig. 3. The oil is delivered to the main crankshaft bearings, from which it finds its way through the hollow crankshaft to the connecting rod bearings. As the oil drips from the connecting rod bearings, it is flung by the revolving crankshaft on to the cylinder walls, the camshaft, and other parts, after which it runs back into the reservoir and is used over again.

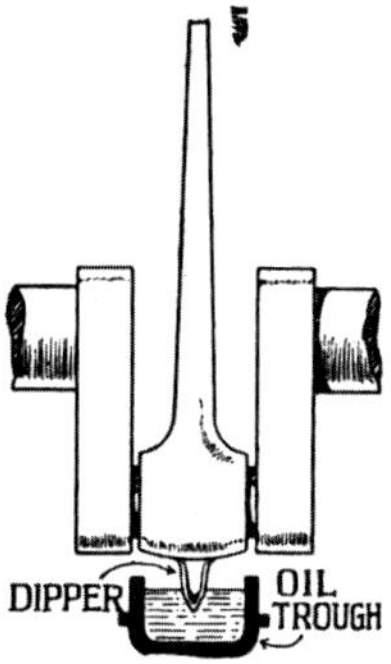

FIG. 4.
The trough system of engine lubrication, where dippers or scoops on the connecting rod ends dip into oil troughs arranged across the crank chamber, and kept filled by means of a pump.

In another system (see fig. 4) the oil is delivered by a pump into narrow troughs, from which a little scoop, or dipper, on the end of the connecting rods, splashes enough oil to lubricate the various bearings, cylinder walls, etc., thoroughly.

In both cases, a float, a gauge glass, or an overflow tap is provided to indicate the quantity of oil in the reservoir, and in many cases the oil is pumped through a glass tube in view of the driver, so that he can tell instantly if the supply fails.

A very important thing to know in regard to a paraffin engine is that a certain amount of paraffin vapour condenses or is never properly evaporated, and finds its way past the piston rings into the oil reservoir. This in time thins the oil to such an extent as to interfere with its lubricating qualities. It should therefore be frequently changed—some makers say every day, some every two or three days. A lot depends on the capacity of the reservoir, but as a general rule a gallon of oil should be used for every eight hours' hard work on a 20 h.p. machine.

To prevent paraffin from getting to the bearings, some makers place the oil reservoir outside the engine. From this a small pump delivers so many drops of oil to each bearing in turn through a system of pipes, a "sight feed" being provided to show that the pump is working properly. After the oil has been used, it drains away through vents on to the ground.

Gear Lubrication.

Gears that are enclosed in an oiltight case should have oil reaching to the underside of the highest shaft on which the pinions are mounted. Gear oil (a thick black oil) should be used, and if this runs out through the bearings to any extent it should be mixed with grease, care being taken not to make the mixture so thick that the pinions merely cut a passage through it.

Greasers.

There are many minor bearings and joints on a tractor which are not advantageously lubricated with oil, and these are provided with screw-down grease cups. A grease not too stiff should be used for these, and cleanliness should be observed in filling them, as a bit of grit or dirt may stop the passage or score a bearing. If one does get stopped up, the stem should be unscrewed and the hole cleaned out at once.

The Cooling System.

Great heat is generated by the explosions in the cylinders, and, though in theory the greater the heat the more power developed, the moving parts affected must be kept sufficiently cool to assure perfect lubrication, and guard against "seizing" through expansion and distortion.

Most motor cycle engines and some aeroplane engines are cooled by so placing the engine that it catches a strong draught of air as it goes along, and the cylinders are covered with a number of thin metal fins to assist radiation in getting the heat away into the air. This method would be

ANOTHER TYPE OF **THE CRAWLEY AGRIMOTOR** AT WORK.

useless on a slow-moving tractor engine; therefore, without exception, tractor engines are water-cooled. The upper parts of the cylinders are surrounded by an enclosed space, termed a water jacket, and the water jacket communicates with a radiator by two pipes, one leading from the top of the water jacket to the top of the radiator, and the other from the bottom of the radiator to the bottom of the water jacket.

The function of the radiator is to keep cooling the water as fast as it gets heated. It consists of a great number of thin perpendicular tubes, often surrounded by radiating fins, and sometimes is corrugated. The water is kept in constant circulation from the cylinder jacket, down through the tubes, and back again, by a pump, and a rapidly revolving fan, placed behind the radiator, draws air between the tubes, thus cooling the water.

The radiator, fan, and pump, being rather expensive items, are sometimes dispensed with, a large tank taking the place of the radiator. Without a pump we have to depend on natural circulation, known as thermo-syphon circulation, so the tank is placed a little higher than the cylinders: and as the water gets hot in the water jacket, being lighter than cold water, it rises through the top pipe into the tank, while cooler water flows along the bottom pipe to take its place in the water jacket. This system of cooling needs five or six times as much water as with a radiator and pump, a 20 h.p. engine requiring about forty gallons, which has to be replenished to the extent of twenty or thirty gallons a day when doing heavy ploughing, whereas two or three gallons would keep a radiator—pump type—replenished. With the tank system the engine also runs much hotter, which makes the constant use of the water drip indispensable to prevent pre-ignition. It might be said in its favour, however, that the engine never attains a dangerous heat if the tank is kept fairly full of water, as the hotter it gets the quicker the circulation. Several different makers employ this system of cooling, and proof that it is successful may be found in the fact that one or two of the most famous tractors are so equipped.

Thermo-syphon circulation is also sometimes employed in conjunction with a radiator, in which case a larger radiator than usual is fitted.

CHAPTER III.

IGNITION.

THE explosive gas is ignited in the cylinders by means of an electric spark, which occurs between two points on the end of a sparking plug. The sparking plug (see fig. 1) is of very simple construction, consisting of a hollow metal body which screws into the cylinder, and a central metal pillar, or electrode, surrounded by insulating material, such as mica or porcelain, which separates the electrode electrically from the metal body. The sparking points are formed by metal projections, which almost form a bridge between the electrode and the metal body, leaving a space of only one-thirty-second of an inch. The electric current generated in the magneto, and conducted to the central electrode, having no further metallic conductor by which to continue its course, leaps the gap in the form of a spark.

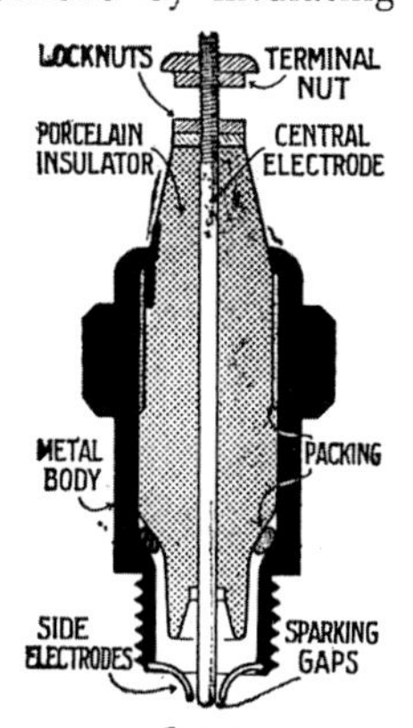

FIG. 1.
Sectional view of a sparking plug.

The Magneto.

Though the magneto is the most delicate and complicated piece of mechanism on a tractor, it needs very little attention, and, in fact, might be considered as reliable as any other part of the machine. Briefly, the usual magneto machine for a four-cylinder engine consists of a reel, or armature, wound with two coils of insulated copper wire, which revolves between the poles of a set of inverted U-shaped magnets, a mechanical interrupting device called a contact breaker, and a distributer, for directing the current to each sparking plug in turn and at the right moment. The contact breaker is fixed to the end of the armature, and the distributer is placed above the contact breaker.

Of the two windings on the armature, one is the primary winding and the other the secondary winding. A current of electricity is generated in the primary winding, which in the first place is not strong enough to leap the gap at the sparking plugs. To intensify this current, advantage is taken of the law of induction, whereby a current of very high voltage is momentarily induced in the secondary winding. To create induction, it is necessary suddenly to break the circuit of the primary current every time a spark is required, and this is brought about by the contact breaker.

To give a more detailed explanation of the magneto would entail a lengthy description of complex electrical phenomena, which would in no way aid anyone in the care

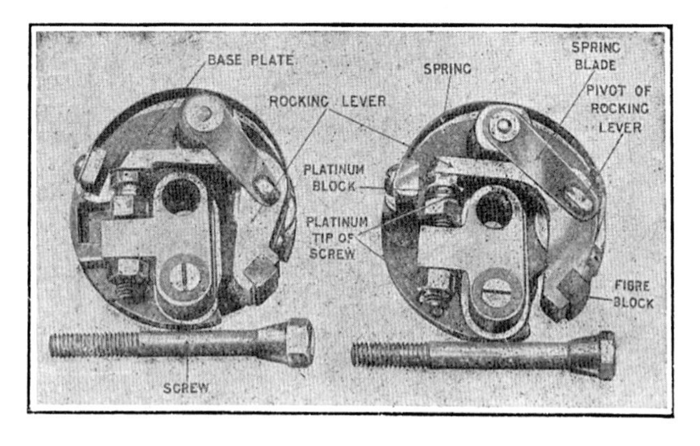

FIG. 2.
Two standard type contact breakers shown removed from the armature ends with which they rotate. The fibre block on end of rocking lever strikes a projection on the adjustable ring shown in the next figure ; by this means the contact points are opened and closed. The long screw shown below each contact breaker passes through the centre hole and fastens the contact breaker to the armature.

and maintenance of a tractor. We will limit ourselves, therefore, chiefly to a description of the parts which are likely to need attention.

The Contact Breaker.

The contact breaker (see fig. 2) comprises a rocking lever pivoted to the circular base plate of the contact breaker, and held upon its pivot by a spring blade. At one end this rocker carries a fibre block, whilst the other end of the rocker, which is L-shaped, has a platinum point, which rests upon another platinum point on the end of an adjustable screw. The two platinum surfaces are held in contact by the bow spring shown in fig. 2. As stated, the complete contact breaker is mounted upon the armature and revolves therewith.

Arranged round the contact breaker is a ring (fig. 3), which is adjustable through a small angle, for a purpose described later, and this ring carries a pair of steel blocks

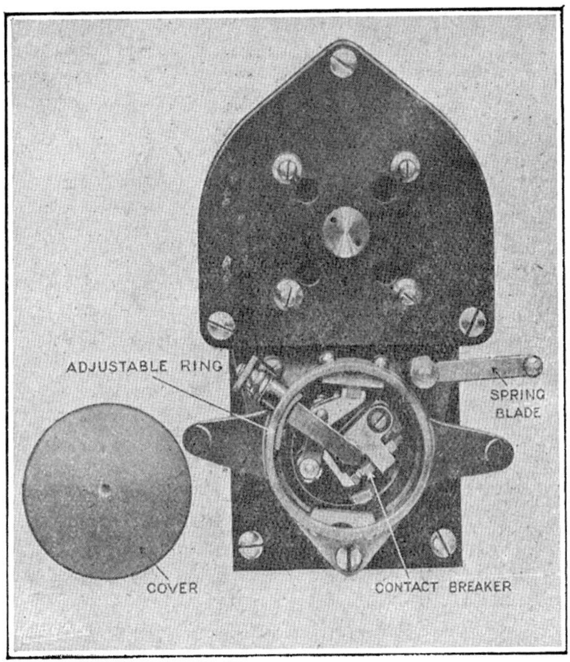

Fig. 3.
End view of a typical four-cylinder magneto machine, showing contact breaker in position on end of armature shaft. When in use the contact breaker mechanism is protected by the cover at left.

called cams (see fig. 4). The ring is stationary in the ordinary way, and so set that, as the armature and contact breaker revolve, the fibre block will strike the cams (as shown in fig. 3) and separate the platinum points. It is already understood that, as the armature revolves, current

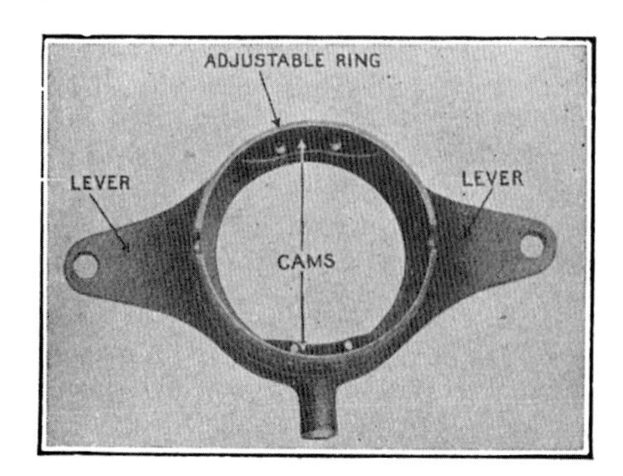

FIG. 4.
Adjustable ring surrounding the contact breaker, and by means of which the ignition point is retarded and advanced. The ring can be rocked through an angle of about 30°. The cams on this model are internal, and one of the levers shown can be connected to a control rod within the driver's reach.

is passing between the platinum points until they are separated, after which a current of very high voltage is induced in the secondary winding. The secondary winding terminates in a slip ring on the armature-shaft.

The Distributer.

The distributer (as shown in fig. 5) comprises a gear wheel, driven by a smaller wheel on the armature-shaft (not shown). Upon the large wheel is mounted a brush holder, suitably insulated. In this is loosely

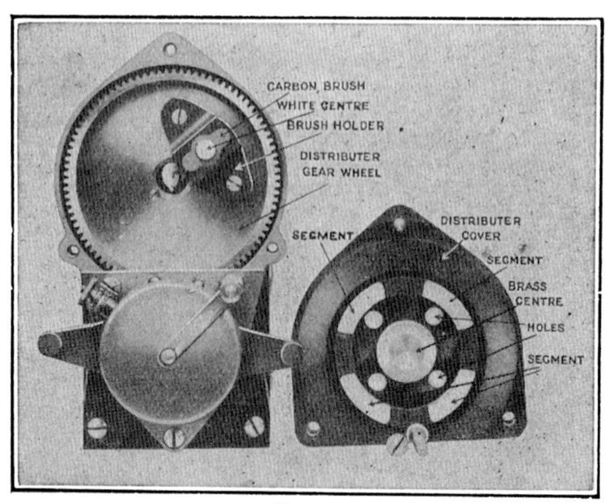

FIG. 5.
End view of a four-cylinder magneto with the distributer cover removed, showing the distribution mechanism by which the current is sent to each of the four sparking plugs in the correct order of firing.

embedded a carbon brush or block (which is a good conductor of electricity) resting upon a spring. To this brush the high-tension current is led by a suitable conductor from the slip ring before mentioned. On the right of fig. 5 is seen the vulcanite distributer cover, which has been removed and placed at the side, and in which are mounted four brass segments. When this cover is placed in position (as it is in fig. 3), the distributer brush will make contact with each of these

THE EAGLE TRACTOR.

ENGINE.	Two cylinders, three sizes: 6×8in. (152 × 203 mm.), 7×8in. (178 × 203 mm.), 8×8in. (203 × 203 mm.).	FUEL.	Paraffin.
		WEIGHT.	No. 1, 43 cwt.; No. 2 45 cwt.; No. 3, 53 cwt.
POWER.	No. 1, 18 h.p.; No. 2, 24 h.p.; No. 3, 32 h.p.	GEARS.	Nos. 1 and 2, one forward, one reverse; No. 3, two forward, one reverse.
CAPACITY.	Three furrows 8in. deep.		

segments in turn, as the wheel on which it is mounted revolves. From the segments the current is conducted directly to the sparking plugs by means of wires attached to the terminal screws (fig. 3), which make contact with the segments through the back of the vulcanite cover.

Timing the Spark.

It was stated earlier in the description of the magneto that the ring carrying the cams (fig. 3) was adjustable through a small angle. This is for the purpose of advancing or retarding the spark. Theoretically, the gas should be ignited when the piston is just commencing the firing stroke. However, it takes an appreciable time for the gas to become ignited and the power of the explosion to be felt on the piston after the spark takes place.

This period is not of much importance when the engine is running slowly, but at high speed it can easily be appreciated that the piston will have moved a long way down on its firing stroke before the gas is fully ignited, unless some means were provided whereby the gas could be fired earlier.

It is therefore necessary for the time of ignition to be variable, being earlier at high engine speeds and later at low speeds. This is effected by moving the adjustable ring by means of a small hand lever. When the ring is moved in the opposite direction to that in which the armature revolves, the fibre block on the rocking arm will strike the cams earlier and an earlier spark will be the result. If it be moved in the reverse direction the spark will occur later. The spark should always be kept advanced as far as possible, and only retarded a little when starting up the engine or when knocking occurs while under load.

Various Types of Magneto.

Though the principles involved are the same in all high-tension ignition systems, the application of them differs considerably in the various types of magneto employed on different tractors. In the K.W. magneto, for instance, the contact breaker does not revolve with the armature, but is stationary and is operated by a cam on the end of the armature spindle, whilst a revolving brass segment makes contact with carbon brushes placed in the distributer cover. Again, the Dixie magneto, which is employed on many tractors, has no armature in the ordinary sense, but a revolving metal structure called a rotor, whilst a primary and secondary coil are wound round a stationary iron core and placed high up in the arch formed by the magnets.

THE EMERSON TRACTOR AT WORK.

ENGINE.	Four cylinders, $4\frac{1}{2} \times 5$in. (114×127 mm.)	GEARS.	Three forward and one reverse; drives are direct at all speeds.
POWER.	12 h.p. (2,400 lb., assuming 200 lb. pull per h.p.), 20 h.p. at belt pulley.	CAPACITY.	Three or four-furrow plough, two self-binders or mowers or full-sized threshing machine.
FUEL.	Paraffin, petrol for starting.	WEIGHT.	49 cwt.

The Impulse Starter.

The magneto on most of the larger tractor engines is fitted with what is known as an "impulse starter," which is a device to aid starting by causing the magneto to give a good spark when the engine is being only slowly turned by hand.

A simple form of this device is shown in fig. 6, and to appreciate its principle it should be understood that the quicker the armature is revolving when the platinum points separate, the more intense will be the spark at the sparking plugs.

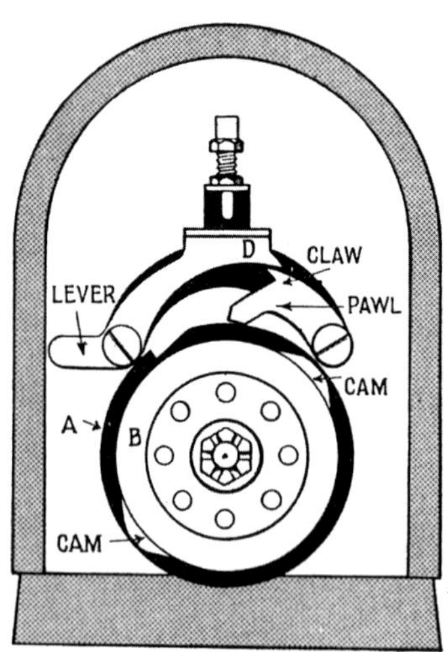

FIG. 6.
Quick-starter device fitted to tractor magnetos to ensure a hot spark at slow speed.
A. Notched disc.
B. Cam disc.
D. Resting point of pawl when out of action.

The black notched disc A is rigidly fixed to the armature spindle on the driven end of the magneto, whilst the white disc B, carrying the cams, is rigidly fixed to the magneto driving-shaft, and the drive is transmitted from B to A by means of a coil spring coupling. In the operation of starting, the pawl or catch engages with the notch in the disc A, holding the armature stationary until the pawl is lifted out of engagement by one of the cams, whereupon the tension of the spring causes the armature to be carried quickly forward, the result being a good spark. This process is repeated until increasing speed causes the cams to strike the pawl with such force as to throw the claw above the point D, where it nominally rests until it is released for starting by pressing the lever.

Care of the Magneto.

For the magneto to function properly, the platinum tips of the contacts (shown in fig. 2) must be clean and

THE BEEMAN GARDEN TRACTOR.

This machine is useful for cultivating small holdings, orchards, and gardens, where a larger machine could not be conveniently used.
It can also be usefully employed as a tractor for hauling a mower or roller, which can be attached to the draw-bar at the rear.

ENGINE.	Single - cylinder, four-stroke, $3\frac{1}{2}$in. × $4\frac{1}{2}$in. (89 × 114 mm.); speed, 230 to 2,000 r.p.m.
POWER.	$1\frac{1}{4}$ h.p. at the draw-bar, 4 h.p. at the belt pulley.
FUEL.	Petrol: one gallon per five hours in the field, and one gallon per seven hours on belt work.
WEIGHT.	4 cwt. 2 qrs.
COOLING.	Water, thermo - syphon, honeycomb radiator, and fan.
LUBRICATION.	Splash.
IGNITION.	Magneto.
OVERALL DIMENSIONS.	Width, 1ft. 5in.; height, 3ft. 3in.

smooth. They must also, when the contact breaker operates, separate to the correct distance. To test this distance, manufacturers supply a gauge, which is sometimes combined with a small pocket-knife and file. Fig. 7 shows a small tool which is supplied with the magneto, the spanner portion being used for unscrewing the central screw, which retains the contact breaker in place, and for undoing the lock-nut for adjustment of the platinum-tipped screw. The gauge indicated is a flat strip of steel, and it should just comfortably pass between the platinum contacts when the fibre block of the rocking lever rests on one of the cams. The distance to which the points should separate is just over one-sixty-fourth of an inch. Sometimes the rocker sticks slightly on its pivot and remains open, with the result that no current is delivered to the sparking plugs. This may occur in damp weather, owing to the fact that the pivot works in a fibre bush, which swells with dampness. To remedy the matter, the curved spring is disconnected at one end, the rocker removed, and the hole eased slightly.

FIG. 7.
Combined gauge, screwdriver, and spanner for magneto machine. The gauge is placed between the contact points, when separated by the cam, to measure the width of the gap.

Should oil get on the platinum surfaces, a piece of paper soaked in petrol, passed between, will clean them.

The contact breaker can be completely removed by undoing the central screw shown in fig. 2.

The only other part which requires attention is the distributer, and as this has a carbon brush, the carbon may scratch the vulcanite plate between the brass segments and leave a line of carbon dust, which, being a conductor, forms a bridge from one segment to another, causing the engine to misfire. From time to time this carbon dust should be removed with a piece of rag moistened with petrol. If the parts mentioned are periodically attended to—say every month when working regularly—then the magneto may be practically ignored in the event of any misfiring or failure to start. In case this latter does occur, attention should first be directed to the sparking plugs. They may fail through soot or oil being deposited on the inner end of the insulation or across the points, in either case forming a bridge by which the current can pass without causing a spark. Moisture on the inside of the sparking plug, or a cracked porcelain, will put a plug out of commission, whilst other parts which may be looked to, if the plugs seem

THE FARMER BOY TRACTOR.

ENGINE.	Four cylinders, $3\frac{3}{4} \times 5\frac{1}{4}$in. (95 × 133 mm.)	FUEL.	Petrol or paraffin.
		WEIGHT.	16 cwt. 3 qr.
POWER.	10-12 h.p. (2,000-2,400 lb., assuming 200 lb. per h.p.), 18-20 h.p. at belt pulley.	GEARS.	One forward and one reverse.
		CAPACITY.	Two or three furrows, equal to six good horses.
WHEELS.	Three. Main driving wheel, 4ft. 2in. × 12in. face; front wheel, 1ft. $10\frac{1}{2}$in. × 6in. face; land wheel, 3ft. 4in. × 6in. face.		

all right, are the wire connections. The air gap, across which the spark jumps, should be about one-thirty-second of an inch.

One or two spare plugs should always be carried, though the modern sparking plug is extremely reliable, and a set will often last for twelve months without giving the slightest trouble.

Oiling the Magneto.

On most magnetos lubricators are provided at each end of the machine, and one or two drops of oil should be given to these about every week, when working regularly. Too much oil is apt to get on the contact breaker, distributer, and armature, which at all times should be kept free from oil.

THE FORDSON TRACTOR DRAWING A PLOUGH.

For Specification see Chapter VII., page 111.

The Fordson Ignition System.

The combinations of a set of magnets, a primary winding, a secondary winding, a contact breaker, and a distributer, which are the essential features of an ordinary magneto, are in the Fordson system split up into various separate units and located on different parts of the engine.

The generator which supplies the low-tension current consists of sixteen horseshoe permanent magnets attached to the flywheel, and revolving close to the same number of bobbins mounted on a plate attached to the cylinder casting. The bobbins are wound with a continuous strip of copper wire, and when the magnets revolve close to the bobbins a low-tension current is generated in the windings. As yet, the current is not powerful enough to cause a spark at the sparking plugs, so it is next distributed to each of four induction coils in turn (one for each cylinder), which are placed in a box on the right-hand side of the engine. Distribution is effected by the commutator placed on the front of the engine and worked by the end of the valve camshaft. It consists of a revolving steel roller, which makes contact with four segments in turn, inside a round case, the segments being connected to the induction coils by wires.

Each induction coil consists of a soft iron core or pillar—running through the centre—around which is wound a primary and a secondary winding, the whole being enclosed in a sealed box. The low-tension current is caused to flow through the primary winding, and a high-tension current is induced in the secondary winding by rapidly interrupting the low-tension circuit by means of a vibrator or trembler. A vibrator (fig. 8) is mounted on each induction coil, and consists briefly of a bridge placed some little distance above the iron core which protrudes, and a flat steel spring blade secured to the body of the coil at one end by screws and provided near the other end with a slight projection, or contact point, placed between the bridge and the core with the contact point nominally in contact with a similar point on the underside of the bridge, the space between the contact points being about one-thirty-second of an inch when the points are separated.

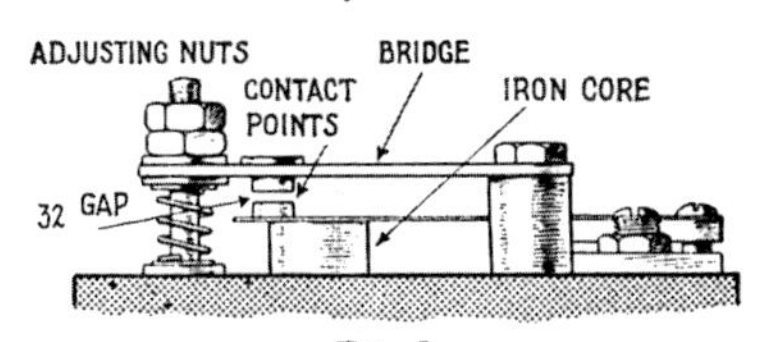

FIG. 8.
Vibrator or trembler fitted to coil box of the Fordson tractor ignition system.

The spring blade and the bridge are so arranged that they form part of the circuit through which the low-tension current must flow. The moment the commutator directs the current through a particular coil, however, the iron core becomes magnetised and attracts the spring blade out of contact with the bridge, thus interrupting the current in the primary winding and inducing a high-tension current in the secondary winding, which is conveyed directly to the sparking plug, with which it is connected by insulated wires.

The moment the low-tension circuit has been broken by the action of the magnetised core on the spring blade, the core becomes demagnetised and the blade springs back to make contact with the bridge, whereupon the current flows once more, the core again becomes magnetised and attracts the spring blade; thus the process of make and break is repeated over and over again at a very rapid rate whenever the distributing roller is passing over any of the segments in the commutator case; therefore, instead of a single spark as given by an ordinary magneto, a very quick succession of sparks occurs during the firing period at each sparking plug, ensuring a very positive ignition of the explosive gas.

To obtain the best results, the contact points of the tremblers must be kept clean and smooth by a very fine file, and the space of one-thirty-second of an inch must be maintained between the contact points. This is done by adjusting nuts which determine the height of the bridge.

It is necessary that the commutator should be well oiled, but at the same time it should at all times be kept free from corroded oil and dirt, so that the roller can make a good clean contact.

CHAPTER IV.

TRANSMISSION AND STEERING.

AS stated in a previous chapter, the speed at which various engines develop their full power varies from 350 to 1,000 r.p.m. It is obvious, therefore, that, as a tractor is necessarily a slow-moving machine, the power must be transmitted to the driving wheels through a reducing gear, capable of giving a great reduction of speed. For instance, the axle speed of a tractor with 5ft. driving wheels,

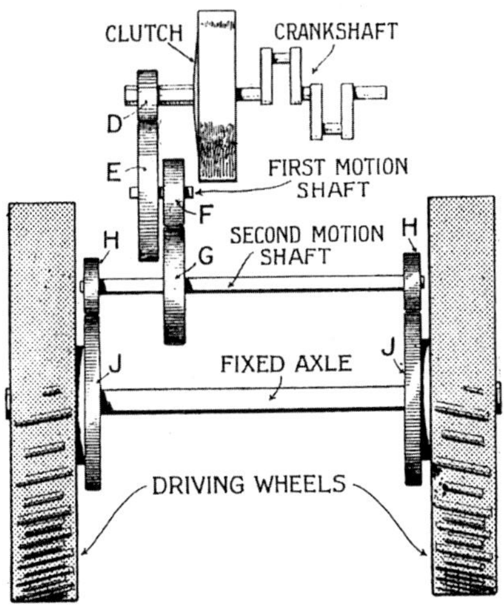

Fig. 1.

Diagrammatic Elevation of the Transmission Gear on the Average Tractor.

D. Pinion on clutchshaft.

E. Gear wheel on first motion shaft.

F. Pinion on first motion shaft.

G. Gear wheel on second motion shaft.

H. Driving pinions on second motion shaft transmitting power to land driving wheels.

J. Driven gears attached to land driving wheels.

travelling at $2\frac{1}{2}$ m.p.h., would be only about 15 r.p.m. If the engine speed was 750 r.p.m., then a reduction or gear ratio of 50 to 1 would be necessary.

The reduction may be obtained by employing a train of toothed or pinion wheels (see fig. 1). A small pinion D on the end of the clutchshaft drives a large pinion E on what is termed the first motion shaft; another small pinion F on the first motion shaft drives a large pinion G on a second motion shaft. The final drive from the second motion shaft to the driving wheels may be by small pinions H H on the ends of the second motion shaft, meshing with larger pinions J J bolted to the driving wheels (as shown in fig. 1), or these pinions may be replaced by sprockets and the final drive be by roller chain. Sometimes, in place of the larger pinions J J internal toothed rings are used, and sometimes roller pinions take the place of the toothed pinions H H; these consist of a number of hardened steel rollers, arranged in a circle between two plates.

The gear ratio obtained, of course, depends on the difference between the number of teeth on the driving and the driven pinions, or chain sprockets, as the case may be.

THE FOWLER MOTOR PLOUGH.

ENGINE.	Two cylinders, 4×5in. (102 × 127 mm.)	CAPACITY.	Two furrows, 3-3½ acres per day; harrow or cultivator may be attached instead of plough frame; can be adapted to drive chaff cutters, grinding mill, etc.
FUEL.	Petrol or benzole.		
GEARS.	Two forward, one reverse; drive to countershaft by worm gearing.	WEIGHT.	22 cwt.

For instance, a twelve-toothed pinion would have to revolve four times to drive a forty-eight-toothed pinion round once.

In addition to the arrangement of pinions shown in fig. 1, a friction clutch is provided for readily disconnecting the engine from the rest of the transmission, so that the engine may be running whilst the tractor is standing still. If the transmission is designed to give more than one forward speed, combinations of pinions are employed which give different ratios. This is called the change-speed gear. In

OTHER VIEWS OF **THE FOWLER MOTOR PLOUGH** AT WORK.

such cases the engine may be started and run with the gear wheels in the neutral position. A reverse gear is also a necessity for driving backwards, and a machine with two driving wheels is usually equipped with a device called the differential gear, to enable the outer driving wheel to travel faster than the other when turning or taking a corner.

The Clutch.

From the engine crankshaft the power is transmitted through the friction clutch—a device which enables the power to be transmitted to, or disconnected from, the remainder

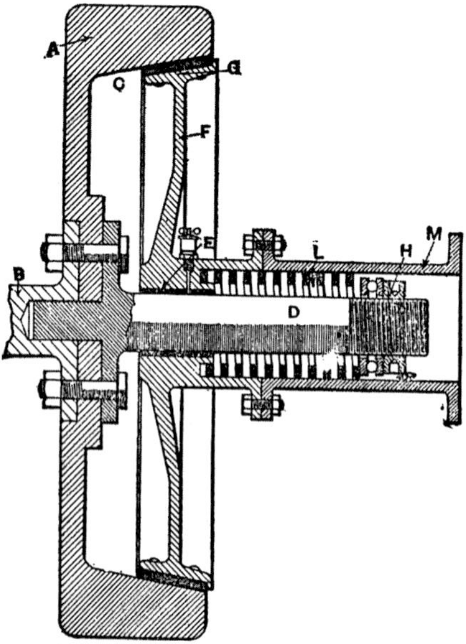

FIG. 2.

SECTION OF AN EXTERNAL LEATHER-LINED CONE FRICTION CLUTCH.

A. Flywheel.
B. Crankshaft.
C. Conical recess in flywheel.
D. Clutchshaft.
E. Spigot bearing.
F. Clutch disc.
G. Conical rim of clutch disc covered with leather.
H. Adjusting nut.
L. Clutch spring.
M. Tubular shaft connecting clutch disc to transmission.

of the transmission gear either rapidly or gradually. The simplest and most usual form of clutch is that shown in fig. 2. This type of clutch is called a cone clutch. The flywheel A, which is attached to the crankshaft B by bolts, is formed with a conical recess C, and has attached to it a short projecting shaft D. On this is a bearing E formed upon the clutch cone F. The conical rim G of this is covered with leather, or frictional fabric. This cone is called the internal clutch member, whilst the conical recess

C in the flywheel constitutes the external member. Between the bearing bush E and a nut H, which is adjustable on the shaft D, is a very strong coil spring L, which always keeps the clutch cone F pressed firmly into the flywheel. The disc F is connected to a tubular shaft M, from which the power is transmitted to the change-speed or reducing gears. When the clutch parts are in the position illustrated, if the flywheel is rotated it is obvious that the clutch disc F will rotate with the flywheel; if, however, the tubular shaft is moved to the right, which is effected by means of a pedal,

VIEW OF **THE HOLT CATERPILLAR** AT WORK.

the spring L is compressed and the conical disc F is drawn out of the flywheel. The flywheel can now be rotated, and the disc remains stationary. As stated, the disc F is connected to the driving wheels through the transmission gear, and consequently the tractor can remain stationary with the engine running.

To start the tractor, the clutch pedal is gently released, allowing the hollow shaft M and the clutch disc F to move slowly towards the flywheel. This is called "letting in the clutch." The leather-covered rim of the disc F gradually comes into contact with the conical surface C of the flywheel, gently taking up the drive and moving the tractor forward.

Another very popular form of clutch is the plate clutch, shown in simple form in fig. 3. Here the flywheel is formed flat on the face A, and it overhangs at B. Fixed to the driven shaft C, which is connected to the gears, is a plate or disc D, which can be pressed up against the surface A by means of a ring E. When the ring is so pressed, it will be seen that the plate D is gripped to the flywheel and must

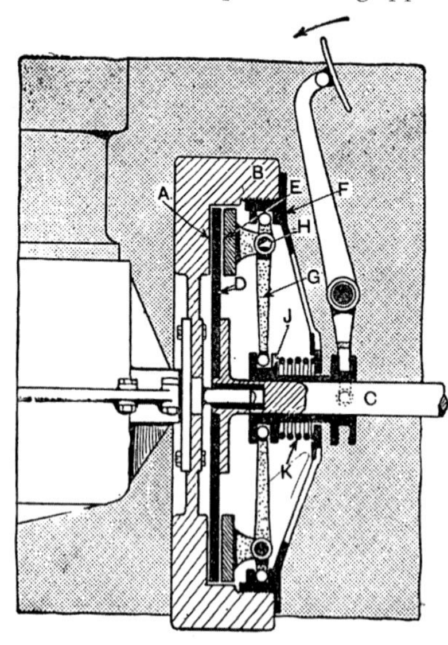

FIG. 3.

A SIMPLE TYPE OF SINGLE-PLATE CLUTCH.

A. Flywheel face
B. Flywheel rim.
C. Driven clutchshaft connected to transmission.
D. Single plate or disc attached to driven shaft C.
E. Ring attached by lugs to levers G.
F. Grooved ring fixed to flywheel.
G Levers pivoted at H.
J. Grooved sleeve operating levers G.
K. Clutch spring.

rotate therewith, in which case the shaft C is clutched to the engine driving-shaft. By moving the ring E to the right into the position shown, the pressure is removed, and the plate D is separated from the flywheel, allowing the former to rotate without turning the shaft C.

The movement of the ring E is effected in the following manner: Screwing into the overhanging part B of the flywheel is a grooved ring F, into the groove of which project the short ends of levers G, pivoted at H to lugs or projections formed on the ring E. The long ends of these levers, of which there are three or more, lie in a groove

THE IDEAL TRACTOR. A BRITISH-MADE MACHINE BUILT AT HEDNESFORD.

ENGINE. Four cylinders, $4\frac{21}{64} \times 5\frac{1}{2}$in. (110 × 140 mm.)
POWER. 35 h.p. at 1,000 r.p.m.
FUEL. Petrol.
WEIGHT. 90 cwt.
GEARS. Two forward, one reverse.
CAPACITY. Four-furrow plough or less according to the nature of the land.
WHEELS. Four. Back, 5ft. 6in. diameter × 10in. face; front, 3ft. diameter × 6in. face.

OVERALL DIMENSIONS. Length with plough, 20ft.; length without plough 10ft.; width 6ft. 6in.; ground clearance, 1ft. 4in.

in a sleeve J, which is moved to the left by a spring K and drawn to the right by moving the clutch pedal in the direction of the arrow. The clutch is generally enclosed, and when all the frictional surfaces are metal the case contains oil.

There are many other forms of clutches, two which are frequently used being the expanding clutch and the contracting clutch. In the former, metal shoes expand inside the hollow flywheel, gripping the internal circumference. In the latter, a metal strap lined with frictional fabric contracts and grips the outside of a drum bolted to the end of the crankshaft or to the flywheel.

THE "INTERSTATE" TRACTOR. FOR MEDIUM-SIZED WELL-APPOINTED FARMS.

ENGINE. Four cylinders.
POWER. 20 h.p. at 850 r.p.m.
FUEL. Paraffin, starting on petrol.
WEIGHT. 50 cwt.
GEARS. Two forward and one reverse, 2 and 3 m.p.h.
WHEELS. Four. Driving wheels, 5ft. 10in. face, fitted with land spuds.
CAPACITY. Three-furrow plough, about 8 acres per day.

The Change-speed Gear.

The great majority of tractors have two forward gears, and a few have three forward gears, usually giving speeds of about two and three, or two, three, and five miles an hour. As stated, these different gear ratios are obtained by employing combinations of pinions, and any of these combinations may be brought into action at will by means of the change-speed lever. The change-speed gears are usually enclosed in an oiltight chamber, termed the gear box. Fig. 4 is a diagrammatic sketch of a two-speed gear cut open. It will be seen that there are two small pinions A and B on the first motion shaft and two larger pinions C and D on the second motion shaft. The pinions A and B may be moved along the shaft to right or left by means of the forked lever E. The pinions, however, can only revolve with the shaft, as the shaft is either square, or formed with grooves or castellations corresponding to similar castellations or squares made in the centre of the pinions. The two larger pinions have no lateral movement. If the forked

FIG. 4.
SECTION OF A CHANGE SPEED GEAR BOX PROVIDING TWO GEAR RATIOS. THE NEUTRAL GEAR POSITION IS SHOWN.

A C. High speed gear wheels. When A is moved to the right it meshes with C, and power is transmitted to the second motion shaft.

B D Low speed gear wheels. Moving B to the left meshes it with D, and power passes to the second motion shaft, which turns at a reduced speed.

Chains convey the power to the land wheels.

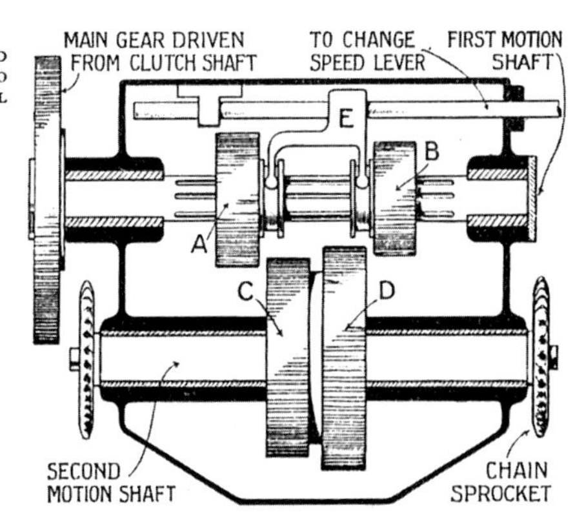

lever E be moved to the right, the pinion A will be brought into mesh with the pinion C. If the forked lever E be moved to the left, then the two pinions B and D will mesh together. B is the smaller of the two driving pinions, and D is the larger of the two driven pinions; therefore, when B and D are in mesh, the combination gives a lower speed than is obtained when the forked lever is moved to the right and the pinions A and C are in mesh.

The position of the gears shown in fig. 4 is known as the neutral position, as neither of the speeds is in gear. When in this position the engine can run without driving the tractor, though the clutch may be engaged. Before attempting to move the gear lever, however, to engage the gears, it is very important first to put out the clutch, otherwise the first motion shaft would be revolving and the second motion shaft would be stationary. An attempt to engage the gears under such conditions would only result in a grinding of the edges of the teeth together, which if persisted in might cause great damage. It is also very necessary to put out the clutch whilst changing from one speed to another, for when passing through the neutral position the second motion shaft must necessarily come to

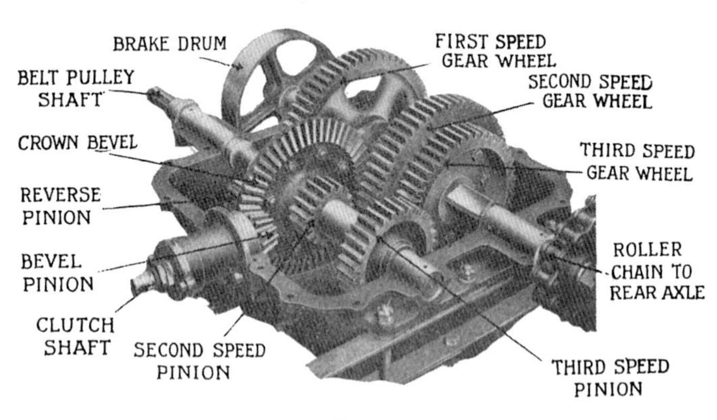

FIG. 5.

Three-speed gear fitted to the Alldays General Purpose Tractor. The gear box is shown with the top half removed; when the box is closed, the wheels run in a bath of lubricant. The power is transmitted to the rear axle by a single heavy roller chain.

a standstill when the tractor stops, but if the clutch were left in, the first motion shaft would continue to revolve.

Figs. 1 and 4 show the arrangement of gears employed when the engine crankshaft is set across the frame and parallel to the rear axle. When the crankshaft is set in line with the frame and at right angles to the rear axle, it will be appreciated that the power must be transmitted to the driving wheels through a right angle. Such a drive may be obtained by employing a bevel pinion and a crown bevel, as shown in fig. 5. Fig. 5 shows a gear box giving three forward speeds. The sliding pinions in this case are on the second motion shaft, which is square, and the final drive to the rear axle is by single roller chain.

The Worm Drive.

A system of transmission is gaining favour whereby right-angle gearing is employed for the final drive. The first motion shaft is set in line with the crankshaft, from which it takes its drive through the clutch; a continuation of the second motion shaft to the rear carries a kind of spiral gear known as a worm, which drives at right angles a much larger gear, termed a worm wheel, placed on the rear axle (see fig. 6). This type of final drive gives a considerable gear reduction in a small space, with the additional

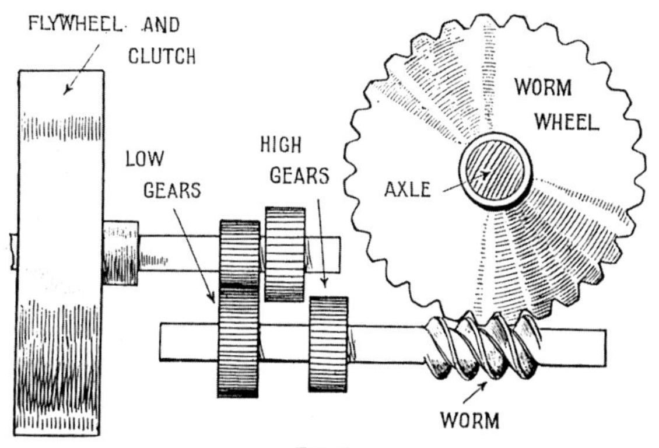

FIG. 6.
Diagrammatic view of a worm and worm-wheel transmission for the final drive to the rear axle.

advantage that it enables the entire transmission mechanism to be enclosed in an oiltight case.

The Friction Gear.

This is a type of change-speed gear coming into use in America and already seen in this country. An example is shown in fig. 7, whereby the drive is transmitted by friction from the edge of a wheel A (covered with frictional fabric) on the engine-shaft to the face of one or other of the large discs B B placed on a countershaft C. The discs B are movable along the countershaft from right to left and *vice versa* by means of a lever, one being to drive the machine forwards and the other to drive it backwards. The friction wheel A, by means of another lever, can be caused to slide along its shaft so that it may press against the face of whichever disc it is engaged with, at any desired point between its outer edge and near its centre. A moment's reflection will show that the disc B would not be driven so fast when the friction wheel A drives it near its outer edge as it would be were it driven at a point nearer its centre. Therefore, by sliding the friction wheel A along its shaft an unlimited number of speeds can be obtained, which can be varied while the machine is running.

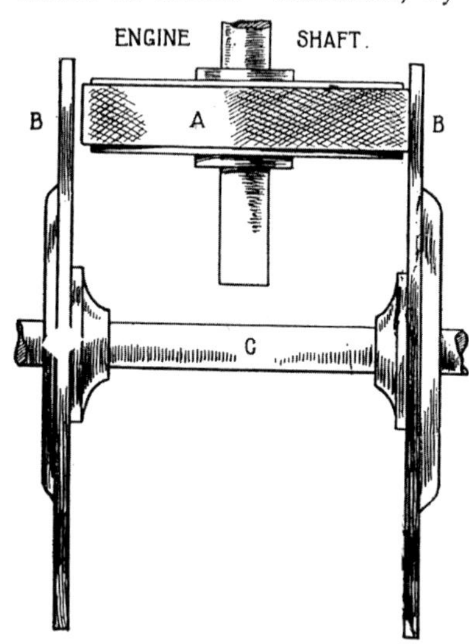

FIG. 7.
Diagrammatic plan of a friction change speed. A, friction wheel. B B, friction discs. C, countershaft.

By sliding the discs B along the countershaft so that the friction wheel A is midway between the two, a neutral position is obtained, and as either disc (unlike toothed gearing) can be pressed into contact with the friction wheel while the engine is running, a clutch in the ordinary sense can be dispensed with. The drive is finally taken from the countershaft C through toothed reducing gearing and a differential to the driving wheels.

The Reverse.

A tractor is usually caused to travel backwards by placing the gears in the neutral position and then bringing an intermediate pinion into mesh with one of the pinions on the first motion shaft and one of the pinions on the second motion shaft. The lowest speed pinions are usually employed for the reverse, as a high reverse gear is not desirable.

THE IVEL TRACTOR.

ENGINE.	Two cylinders, 6¼ × 6in. (159 × 152 mm.)	CAPACITY.	Three-furrow plough, cultivator, reaper, and mower.
POWER.	13 h.p. (2,600 lb., assuming 200 lb. per h.p.), 24 h.p. at belt pulley.	WHEELS.	Three. Driving wheels, 40in. diameter × 10in. face; steering wheel, 20in. diameter × 10in. face.
FUEL.	Paraffin.		
WEIGHT.	37 cwt.	OVERALL DIMENSIONS.	Length 9ft., width 5ft. 7in., height 5ft. 6in.
GEARS.	One forward, one reverse.		

The Differential.

The action of a differential gear is rather difficult to understand without a model, and still more difficult to explain by means of a description and drawing. Its construction is not important to the average tractor owner, and if he so desire he may ignore it and omit the following technical description.

Fig. 8 shows a very elementary form of differential gear. The two pinions C D of the change-speed gear (fig. 4) are here shown in section, exposing the interior. The second motion shaft A A, it will be observed, is in two halves, and the gear wheels C D are not fixed to the second motion shaft, but are free to revolve thereon. On the inner end of each half of the second motion shaft — frequently called the differential-shaft—bevel pinions E E are rigidly fixed, whilst two other bevel pinions F F, free to revolve on spindles carried by the gear wheels C D, mesh with the two bevel pinions E E. Thus, whilst the gear wheels C D are free on the shafts A A, the shafts A A are driven by the gear wheels C D through the agency of the four bevel pinions.

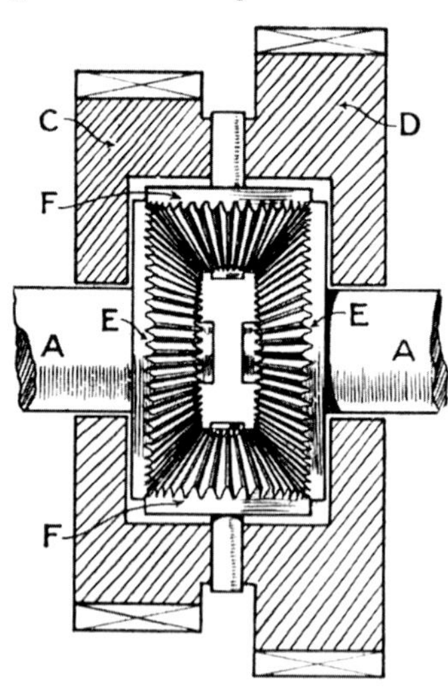

FIG. 8.
SECTIONAL VIEW OF A BEVEL PINION TYPE DIFFERENTIAL OR BALANCE GEAR.
A A. Divided axle.
C D. Change speed pinions.
E E. Bevel wheels attached to ends of divided axle.
F F. Bevel pinions free to rotate on their own shafts, which are carried by the change speed pinions C D.

When the vehicle is travelling in a straight line, the bevel pinions E E, F F, the gear wheels C D, and the shafts A A turn round solidly as though in one piece. If, however, the vehicle is travelling in a circular direction, the two shafts must travel at different speeds, as the outside driving wheel describes a larger circle than the inside one. When the inside driving wheel lags behind, the bevel pinions F F are caused to revolve, allowing the shaft driving the inside wheel to slow down and the other shaft to be hurried forward, so to speak. Notwithstanding the difference in speed of the two shafts, the power from the engine is always proportionately transmitted to each driving wheel. For convenience, only two bevel pinions F F are shown between

THE IVEL-HART TRACTOR.

ENGINE.	Two cylinders, $5\frac{1}{2} \times 7$in. (140 × 178 mm.)	GEARS.	Two forward, two reverse.
POWER.	15 h.p. (3,000 lb., assuming 200 lb. per h.p.), 22 h.p. at belt pulley.	CAPACITY.	Three-furrow plough, cultivator, reaper, and mower.
FUEL.	Paraffin.	WHEELS.	Three. One driving wheel, 5ft. 4in. diameter × 2ft. 2in. face; two steering wheels, 2ft. 9in. × 6in. face.
WEIGHT.	59 cwt.		
OVERALL DIMENSIONS.	Length 12ft. 8in., width 7ft. 6in., height 7ft. (to top of stack).		

the bevel pinions E E; in practice, three or four are employed, placed an equal distance apart, to give greater strength and rigidity. When the final transmission to the driving wheels is by a single gear wheel, or by worm drive, or by a single roller chain, as illustrated in fig. 5, the differential is, of course, on the rear axle.

The differential gear has been discarded by one or two makers, and in its stead each driving wheel takes its drive from the transmission gear through a separate clutch. When turning, one wheel is put out of action and the other takes the whole of the drive; the left wheel driving when turning to the right, and *vice versa.* The objection to this method is that, when turning on the headland (which is usually the most slippery part of a field), one wheel being out of action, the tractive grip of the machine is reduced by half. On the other hand, when travelling straight with both wheels

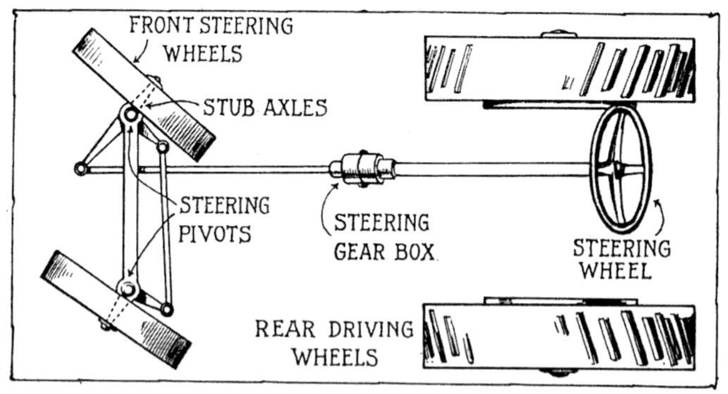

Fig. 9.
Diagrammatic view of the Ackermann type steering gear.

taking the drive, should one wheel encounter a soft place it cannot revolve without the other, as so often happens where a differential gear is fitted without the addition of any locking device.

Steering Gear.

This can be divided into two distinct classes—(1) hand steering and (2) mechanical steering. Hand steering is effected by a hand wheel which can be made to guide the machine by means of a steering column, a worm and pinion and suitable rods or chains acting on the front wheel or wheels.

The tendency of design in hand steering is chiefly in two directions. The motor car, or Ackermann, type (fig. 9)

ANOTHER VIEW OF **THE IVEL-HART TRACTOR** AT WORK.

is usually to be found on the lighter type of tractor, whilst the chain barrel or traction engine type (fig. 10) is generally employed on the heavier machines. With the Ackermann type of steering, the wheels are mounted on short stub axles pivoted on the ends of the main axle. With the chain barrel type the wheels are mounted on the main axle, which is pivoted in the centre.

Mechanical Steering.

Most double chain track and some two and three-wheeled tractors are steered mechanically by a braking arrangement which acts on the shaft on which the differential is mounted.

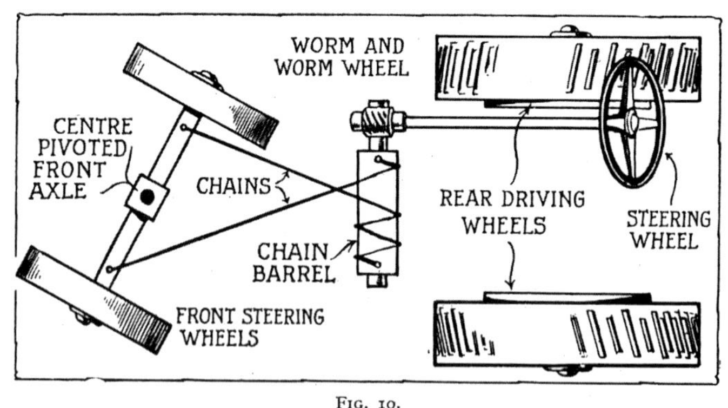

FIG. 10.
Diagrammatic view of centre pivot steering operated by chains and a chain barrel.

After a study of the differential (fig. 8), it will be obvious to the reader that when one of the shafts A A is gripped with a brake the pinions F F will come into play, causing the other shaft only to be driven, which has the effect of allowing the wheel or chain track driven by that shaft to describe a circle, with the other wheel or chain track as its centre. With this type of steering it is possible to turn in a very small radius, which is very convenient in small irregular fields. An ordinary hand wheel and steering column is employed, which, when turned to right or left, tightens a brake on one or the other of the driving shafts.

The steering gear is probably the weakest part on the majority of tractors imported into this country from America. Designed for service on the large stretches of land to be found in America and Canada, they cannot be expected to bear the strain of frequent turning involved when employed in our small fields, which are mere allotments compared with the vast stretches over there.

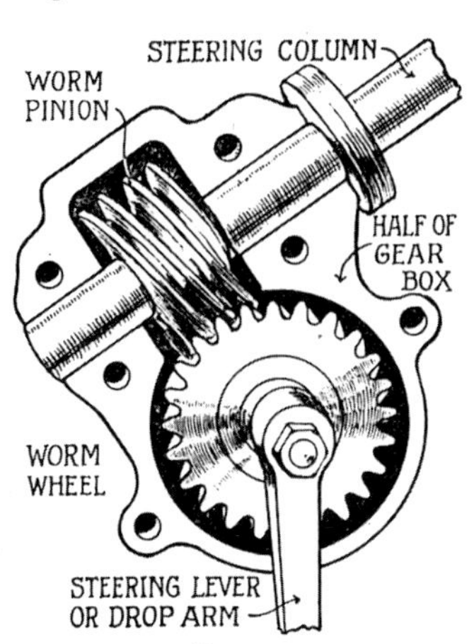

FIG. 11.
Steering gear box with half the case removed exposing the worm and worm wheel.

However, a lot can be done to preserve the conditions of the steering gear by keeping every part well oiled or greased, for stiffness in one part of the gear will react and cause wear in every other part. The worm and pinion (fig. 11) is the reducing gear, which makes it possible to turn a heavy machine by hand. This in the Ackermann steering is usually enclosed in a box (the steering box), which should be filled with soft grease about every two months, or else constantly oiled.

CHAPTER V.

STARTING AND REPAIRS.

THE reader already understands that before the engine can be started it must be first set in motion by means of the starting handle, to suck in the necessary charge of gas and generate an electric spark to ignite it.

Before attempting to start, certain preparations must be made. First of all, there must be petrol in the petrol tank and paraffin in the paraffin tank. There must be oil in the lubricator, or, in the case of it being contained in the crank chamber, it must be ascertained that the right amount is contained therein, for which purpose a gauge glass or overflow tap is provided. The ignition switch (if one is fitted) should now be turned on, and, after making sure that the float chamber does not contain paraffin, the petrol valve may be turned on to the carburetter.

If a needle valve is provided to vary the flow of fuel through the jet or spray, this should be adjusted to give the right amount. (Very often the correct setting is indicated by a pointer or mark.) The needle should be a little wider open for starting than for actual running, and a little wider for paraffin than petrol. See that the ignition lever is retarded a little, as a precaution against backfire, and set the throttle about a quarter open. Finally, a few drops of petrol should be poured directly into each cylinder through the compression taps, or through the hole or tap provided for the purpose in the induction pipe. This ensures a rich mixture, which it is not always possible to get from the carburetter while the engine is being slowly turned by hand.

One or two smart turns of the starting handle should now suffice to start the engine. This accomplished, the ignition lever should be fully advanced, and the fuel valve turned off as far as it will go without interfering with the even running of the engine.

Large engines, and especially the horizontal single and double-cylinder types, are not easy to turn quickly. To most of these is fitted a clever little device known as an impulse starter, which considerably aids starting by causing a good spark when turning very slowly (see Chapter III., page 48). A little lever or trigger is moved by hand to bring it into operation, and this must not be neglected when preparing to start.

To reduce the compression and make starting easier, many of the larger engines are also fitted with a half-

THE 30 H.P. KILLEN-STRAIT. AN ALL-PURPOSE ENDLESS-CHAIN TRACTOR.

ENGINE.	Four cylinders, 4 × 6in. and $4\frac{3}{4} \times 6\frac{1}{4}$in. (102 × 152 mm. and 121 × 159 mm.)	TRACKS.	Endless driving chain 12ft. 10in. long × 18in. wide, endless steering chain 7ft. 6in. long × 13in. wide; load on land less than 4 lb. per sq. in.
POWER.	16 h.p. (3,200 lb., assuming 200 lb. pull per h.p.), 25 h.p. at belt pulley.	CAPACITY.	Four to five furrows in medium land, 7-10 acres per day; all ordinary farmwork; haul 5 tons on good roads.
FUEL.	Paraffin, starting on petrol.		
WEIGHT.	52 cwt.		
GEARS.	One forward and reverse.		

OVERALL DIMENSIONS. Length 12ft. 10in., width 5ft. 10in.

compression release, or decompressor, which consists of a tap placed half-way along each cylinder wall. This tap is opened previous to starting, and as the piston comes up on the compression stroke the gas is expelled until the hole in the tap is covered by the piston, the remaining charge resulting in lower compression.

Work may be commenced the moment the engine is started. As soon as the vaporiser is warm enough, the petrol should be turned off and the paraffin turned on; if petrol economy is a consideration—as it must be at the time of writing—the operator must not forget this, or the reserve supply of petrol in the small tank will quickly be used up, and he will be apprised of his neglect by the engine coming to a standstill. The time taken to warm the vaporiser varies from two to ten minutes with different machines, and is prolonged somewhat in cold weather. The amount of petrol used varies from half a pint to a quart.

Overhauling and Adjustments.

As time goes on, the necessity for overhauling and adjustments will arise in the form of leaky valves, carbonised combustion spaces, and pistons, worn bearings, gear wheels, piston rings, etc. To a great extent the frequency of these will depend on the care with which the tractor is managed, and the cost of putting things right will depend upon how much the owner or his driver can undertake.

As prevention is better than cure, the average farmer will be well advised to do his utmost to prevent the necessity for repairs as long as possible by learning all he can about his machine, and giving it the best attention in his power. Then, when the inevitable wear and breakages do occur, he can call in the aid of a skilled man, knowing that the machine has already paid the bill in the form of services rendered. There are, of course, many simple adjustments which any owner or driver should learn to do himself. Some of these are mentioned under the headings "The Carburetter" (Chapter II.) and "The Magneto" (Chapter III.), and a few others are mentioned below.

Grinding in the Valves.

The constant passage of hot gases through the exhaust valve ports will in time impair the valve faces and seatings, causing leaky compression with consequent loss of power. Loss of compression is evident if the starting handle does not spring back after a piston has been held on full compression a few seconds. The remedy is to take out the valves and grind them in. To take out the valves, remove the springs and unscrew the valve caps which give access to the heads of the valves. If valve caps are not provided, the cylinder heads will be found to be removable in one piece, bringing the valves away with them.

Make a paste with some very fine emery or carborundum powder and thin oil, or, better still, buy a tin of valve-grinding paste from any motor accessory stores. Apply a little of this to the valve face and seat, and revolve it with a screwdriver, or a brace fitted with a screwdriver bit. Take a few turns in one direction and then the other, lifting the valve off its seating occasionally to let the paste get between the valve and seat. Continue until both valve and seat show a smooth bright surface all round, and then wash off the paste thoroughly, being careful that no trace of grit is left.

THE LIGHTFOOT TRACTOR.

ENGINE.	Four cylinders, $3\frac{1}{2} \times 4$in. (89 × 102 mm.)	FUEL.	Petrol or paraffin.
		WEIGHT.	25 cwt.
POWER.	6 h.p. (1,200 lb., assuming 200 lb. pull per h.p.), 10 h.p. at belt pulley.	GEARS.	One forward and one reverse.
		CAPACITY.	Two 14in. furrows.
		TRACKS.	Two, 2ft. 9in. × 6in.

OVERALL DIMENSIONS. Length 8ft. 6in., width 3ft., height 4ft. 6in.

Should the cylinder heads be removed, make sure that a really good joint is made between the heads and the cylinders when reassembling, otherwise the water will leak from the jackets to the cylinders. The makers supply a special joint for this purpose (called a "gasket," and usually made of thin copper and asbestos), with holes already stamped out to correspond with the studs and water ports. One of these should be always kept in stock, so that if the old one is at all doubtful it can be replaced by the new one.

Decarbonising.

Occasionally it is necessary to remove the carbon deposit which forms on the piston heads and combustion chamber walls, as an excess of deposit causes knocking and overheating.

To reach the carbon, in the case of cylinders cast in one piece, the cylinders must be detached from the crank chamber and slid off over the pistons, whilst with detachable heads the carbon is exposed by removing the heads in the

THE LITTLE GIANT TRACTOR.

ENGINE.	Four cylinders, $4\frac{1}{4} \times 5$in. (108 × 127 mm.)	CAPACITY.	Four-furrow plough, hauling reapers, threshing, chaff cutting, and hauling a 6 ton load on the road at 6 m.p.h.
POWER.	16 h.p. (3,200 lb., assuming 200 lb. pull per h.p.), 22 h.p. at belt pulley.		
FUEL.	Petrol, paraffin, alcohol, or naphtha.	WHEELS.	Four. Driving wheels, 4ft. 6in. dia. × 1ft. 2in. face.
WEIGHT.	$46\frac{1}{2}$ cwt.	DIMENSIONS.	Wheelbase 7ft. 3in., track 4ft. 4in.
GEARS.	Three forward, one reverse.		

same manner as described for getting to the valves. Of course, all fittings such as the carburetter, inlet and exhaust pipes, and ignition wires which are fixed to the cylinders or cylinder heads, are dismantled first.

Decarbonising by Oxygen.

The oxygen process of removing carbon deposit from the combustion spaces has been successfully used in motor car practice for many years. No dismantling is necessary for this operation.

A stream of oxygen from a special cylinder is directed through the valve cap or sparking plug aperture, a lighted taper is applied, and the carbon disappears in the form of carbon dioxide gas. Most large garages keep the necessary outfit for this process.

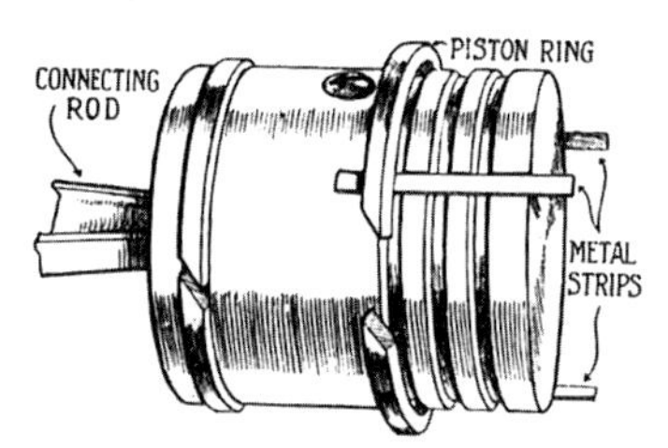

FIG. 1.
Strips of metal placed between the p ston and its rings to facilitate removal of the rings.

Piston Rings.

Sticking piston rings are sometimes the cause of loss of compression, and it is a good plan to remove these from the pistons and scrape out the grooves when the cylinders are off for decarbonising or grinding in the valves. Piston rings, being made of cast iron, are very fragile, and care should be taken when handling them.

The best means of getting them off and replacing them is to slide them over three thin metal strips, as illustrated in fig. 1. When replacing the cylinders over the pistons, it will often be found difficult (especially with the large two-cylinder engines) to get the piston rings to enter the cylinder bore owing to their springy nature. A useful clamp that will overcome the trouble can be made from thin sheet metal and a short bolt (see fig. 2). The clamp should be drawn only moderately tightly around the rings by screwing up the nut. As the cylinder is placed over the piston the clamp will slide forward, not releasing the rings, however, until each one has found its place within the cylinder bore.

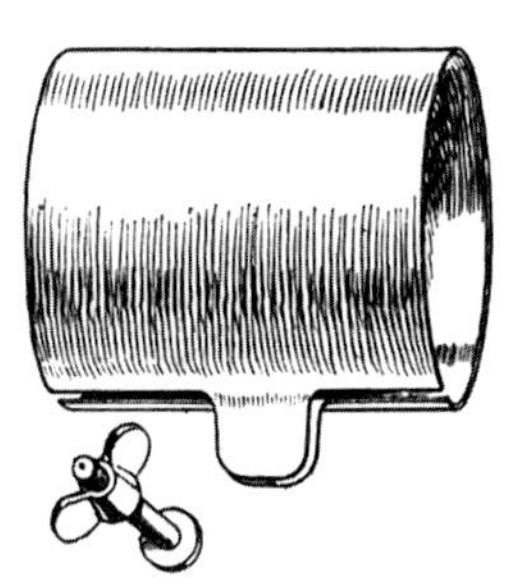

FIG. 2.
A broad clamp made to fit over the piston rings. When contracted by screwing up the wing nut, the clamp enables the piston, with its rings, to be inserted in the cylinder.

Removing the cylinder heads only does not expose the sides of the pistons; therefore, when it is desired to get to the rings, the cylinders must either be removed bodily, or else the connecting rods must be unbolted from the crankshaft and the pistons be pushed out through the top of the cylinders after the heads have been removed. The latter method will usually be found the most convenient when dealing with the larger two-cylinder horizontal engines. As a rule the above-mentioned operations should not require doing more than once in every six months, but it will be found that the amount of carbon deposit formed varies considerably with different engines. Those with the valves placed in the cylinder heads, or with one valve on either side of the cylinder, keep cleaner than those that have the valves all on one side, and some of these latter may need cleaning more often.

It is a good plan to arrange periodically for a few days in and do these jobs altogether. The wisdom of this will be obvious when it is remembered that the dismantling and reassembling of the various parts concerned is a longer job than the actual grinding in, cleaning, etc.

Adjusting the Valve Tappets.

There should be a small clearance between the ends of the tappet rods and the valve stems (about one-fiftieth part of an inch) to allow for expansion and assure the positive closing of the valves. For the purpose of maintaining this clearance, the ends of the tappet rods are adjustable by means of a small cap and lock-nut. These should be checked from time to time and set to give the right clearance.

The Fan Belt.

This should always be kept reasonably tight, as a slipping fan belt is often the unsuspected cause of an overheated engine. Though a fan requires little power to drive it slowly, the resistance offered to the blades by the air at high speed is very great : hence the necessity of keeping the belt tighter than would at first seem necessary.

CHAPTER VI.

HORSE-POWER AND DRAW-BAR PULL.

BEFORE the owner, or prospective owner, of a farm tractor can even form a rough estimate of the ability of a machine, he must know something of "horse-power" (h.p.) and the relation horse-power has to draw-bar pull.

The only horse-power that interests us here is "brake-horse-power" (b.h.p.) Brake-horse-power is the actual horse-power available at the flywheel, and is measured by applying a suitable form of brake to the engine flywheel, and noting the power absorbed by this brake when the engine is running at its full normal speed.

Now work is done when a force is exerted over a measurable distance. Force is measured in pounds, and distance in feet. If a weight of 1 lb. is raised through 1ft., then 1 ft.-lb. of work has been done. Similarly, 10 ft.-lb. would be the work done in raising 1 lb. through 10ft., 10 lb. through 1ft., or 5 lb. through 2ft.

Power is the capacity to do work. It takes twice as much power to raise a weight 2ft. in one second as to raise it 1ft. in the same time, or to raise it 2ft. in two seconds. For estimating the utility of machinery, we want, then, to establish a unit of power. For this purpose the horse-power has been chosen, and is stated to be the power required to do 33,000 ft.-lb. of work in one minute. This was originally supposed to be within the ability of a strong horse, but it has been found that the estimate was too high. Few horses can do more than 25,000 ft.-lb. a minute, and 22,000 is more nearly the average. It would appear, then, that an engine of 1 h.p. can do more work than a horse, and under some circumstances this is so.

In estimating the abilities of a tractor, we must remember that some of the engine power represented at the flywheel and available for belt work is lost in the friction of the transmission mechanism, and some is used in propelling the tractor itself, and is not therefore available for useful work at the draw-bar. It is easy to reckon out that one mile an hour is the same thing as 88ft. per minute. If a tractor moves at five miles an hour it covers 5 × 88 = 440ft. per minute. If, while doing so, there are 10 h.p. available for useful work, the useful work that can be done in the minute is 10 × 33,000 = 330,000 ft.-lb. If we divide this figure by the distance in feet (440), we get the pull in lb. exerted by the tractor, thus—

$$\frac{330,000}{440} = 750 \text{ lb.}$$

This would be the draw-bar pull of the tractor; that is to say, the pull it can exert, through the draw-bar or other connection, on the waggon or implement hauled behind it. If, with the same available horse-power, the speed were increased, the draw-bar pull would be lowered. If the speed were lowered the pull would be increased. It is useless, then, to be told the draw-bar pull unless we are also told the speed at which the pull is given.

Speed and Draw-bar Pull.

Provided we know the speed, it is much more useful to us to be told the draw-bar pull than to be told the horse-power. It really affords a more convenient method of comparison of the engine and the horse. An authority has stated that a farm horse can usually give an average draw-bar pull, throughout the day, equal to one-twelfth of his own weight, at a speed of two and a half miles an hour. As a matter of interest, let us see what this means in horse-power. Say the horse weighs 1,200 lb. The pull will then be 100 lb. He travels at 2½ m.p.h., or 220ft. per minute. In the minute he does 220 × 100 = 22,000 ft.-lb. of work, which agrees with the estimate already given. A heavier horse, of course, would do something more nearly approaching a horse-power; and even a light horse, in starting, or for quite short periods, can exceed a horse-power by a good margin, probably developing, for the moment, 3 h.p. or 4 h.p. As stated, some of the power of a tractor engine is absorbed by friction in the transmission gear, and some is used to propel the machine along. This is why it takes far more than a 2 h.p. tractor to do the job of two horses.

Now it is not much use to be told that a tractor will draw a three-furrow plough, even if the depth and width of furrow are given—which is often not the case—unless we know exactly on what land and on what gradients the work is done, and the condition of the ground at the time. Nor is it very satisfactory to be told the horse-power available at the draw-bar. This must vary immensely, as much more power is used in propelling the tractor itself on soft than on hard ground. Consequently, none of the figures given to us by the manufacturer can be taken as anything more than a rough guide, but at all events they are something to go by. (In the specifications of various tractors given under the illustrations in this book the draw-bar pull has been assumed to be 200 lb. pull per h.p.)

Turning particularly to the work of ploughing, a writer in *The Autocar* has given the following as the results of carefully-conducted tests carried out in England in the autumn of 1916:

Nature of soil.	1 lb. pull required per sq. in. of section of furrow turned.
Sandy	3
Wheat stubble	5 to 7
Clover stubble	8 to 12
Heavy clay	20

Some other tests, also of English origin, give the following figures:

Loamy sand	5
Sandy loam	$5\frac{1}{2}$
Strong loam	10
Blue clay	15

A large number of tests in varying soils in Missouri gave an average of $5\frac{1}{4}$ lb. per sq. in.; in clover, $6\frac{1}{2}$ lb.; and in oat stubble, somewhat under 5 lb. American tests show also that a pull of 14 to 16 lb. per sq. in. is required for ploughing in virgin soil.

As a rough average, not to be applied indiscriminately to every instance, we may put the pull necessary to turn a furrow 6 × 10in. at 375 lb., with a minimum of about 200 lb. and a maximum up to 900 lb. This is the same thing as putting the average at $6\frac{1}{4}$ lb. per sq. in. of the section of the furrow.

To take one example of the use of the figures tabulated above, we may reckon what is needed to plough three furrows, 6 × 10in., in clover stubble, the average for which is 10 lb. per sq. in. The section of each cut is 60 sq. in., giving a total of 180 sq. in. for the three. At 10 lb. per sq. in. this means a pull of 1,800 lb. If the draw-bar horse-power is 10 h.p. we can do 330,000 ft.-lb. per minute, or 330,000 × 60 ft.-lb. per hour. Dividing by the pull (1,800 lb.), we find that we ought to be able to travel at $\frac{330,000 \times 60}{1,800}$ feet per hour, or $\frac{330,000 \times 60}{1,800 \times 1,760 \times 3}$ miles per hour, which equals $\frac{100}{48}$, or just over two miles per hour.

Now if, in our opinion, better ploughing would result from higher speed, we have the alternative of ploughing only two furrows at three miles an hour, which will give almost the same acreage for the day's work as three furrows at two miles. The only difference would be that there would be more turns at headlands, which would involve some loss of time. The choice would depend partly on the land speeds corresponding to the use of the various gear ratios at normal engine speed.

THE MARSHALL TRACTOR.

Capacity of 20 h.p. Tractor.

These calculations roughly confirm the general estimate to the effect that a tractor of 20 to 25 h.p. (say 10 to $12\frac{1}{2}$ h.p. at the draw-bar) will plough three furrows of reasonable depth in anything but very heavy soil. Failure to attain to this standard generally means either inferior design and construction, indifferent running of a mechanism through wear or faulty adjustment, or lack of suitable arrangements to ensure good adhesion. It may also be due to faulty attachment of the plough to the tractor, causing the pull to be taken at the wrong point.

All our estimates up to this point, of course, will be falsified if the ground is extremely hilly. The authority already quoted from *The Autocar* states that he can corroborate the following theory: For every 1% of gradient, deduct from the effective draw-bar pull $1\frac{1}{2}$% of the total weight of tractor and plough. Thus, with a tractor and plough weighing 50 cwt. (5,600 lb.) and having a draw-bar pull of 1,800 lb., we must for a 5% gradient (1 in 20) deduct $\frac{5 \times 1\frac{1}{2} \times 5{,}600}{100}$ lb. = 420 lb., reducing the draw-bar pull at normal speed to 1,380 lb. If the weight of tractor and plough, for the same draw-bar pull, could be halved, then the reduction for a 5% gradient would be only 210 lb. An American authority states that the power actually employed in propelling a tractor up a gradient can be calculated on the following basis: The effort exerted (in pounds) is equal to about 1% of the weight of the tractor for every 1% of gradient. The tractive efforts necessary to move a power-propelled machine on various classes of surface are given as follow:

	Lb. per ton weight.
On rails	5.16
On asphalt or hard wood	12.24
On macadam	30.60
On loose gravel	150-200
On sand	400

Take a tractor weighing two tons and hauling four tons, giving a total weight of six tons. On level macadam this would involve a pull of about 6 × 30 = 180 lb. On an up grade of 10% (1 in 10) we should have to add 10% of the weight, or $\frac{10}{100} \times 6 \times 2{,}240$ lb. = 1,344 lb., giving a total pull of 180 + 1,344 = 1,524 lb. At five miles per hour, or 5 × 88ft. per min., the h.p. would be $\frac{5 \times 88 \times 1{,}524}{33{,}000}$ = about 20 h.p. From the above we see that a suitably-designed tractor, capable of ploughing in three furrows in average soil, should have no difficulty in handling, say, four tons of useful load on any ordinary road surface, travelling at 5 m.p.h. on the level and at about 2 to 2½ m.p.h. on steep gradients. Its suitability for such work is not a question of power, but of adhesion, gears, springs, brakes, and other factors. Directly it diverges from the road on to loose sand or soft ground, the work of haulage becomes immensely more arduous, and such considerations as wheel diameter and width, and the provision of spuds or spikes to help adhesion, take on an added importance.

To sum up, it is most difficult, if not impossible, for the tractor manufacturer to give any exact figures which will really enable the prospective purchaser to estimate accurately in advance the capabilities of a tractor. Probably the most useful is a statement of the draw-bar pull given on a level surface at some stated speed, which would preferably be the normal speed at which the tractor is intended to plough.

Influence of Diameter and Width of Wheel on Soil Compression and Draw-bar Pull.

On soft or loose ground, both soil compression and the draw-bar pull are considerably affected by the diameter and width of the driving wheels of a tractor. In other words, unsuitable wheels will not only greatly increase "packing" of the soil, but with such wheels more power will be absorbed in propelling the machine, thus reducing the power available for useful work at the draw-bar.

First of all, we will consider the influence of wheel diameter on soil compression, and for comparison we will imagine we have two tractors of equal weight standing side by side on soft ground. All four driving wheels are 10in. wide, but those on one machine are double the diameter of those on the other.

MARTIN'S SELF-CONTAINED THREE-FURROW MOTOR PLOUGH AND AGRICULTURAL TRACTOR.

FITTED WITH ENDLESS TRACK CREEPERS.

ENGINE.	Four cylinders, $3\frac{3}{4} \times 5$in. (95 × 127 mm.)	GEARS.	One forward and one reverse.
FUEL.	Petrol.	CAPACITY.	Three-furrow plough, cultivating, harrowing, seeding, binding, mowing, etc.
WEIGHT.	35 cwt.		

The ground being soft, the wheels will not stand on the surface, but will sink in until a sufficiently large area of ground is in contact with the wheel treads to support the weight of the vehicle. Suppose it requires 200 sq. in. of ground surface to support the weight on each wheel, the wheels will sink in until each has formed a depression about 10 × 20in. Reference to fig. 1 will show that, in seeking an area of contact sufficient to support the weight, the smaller wheels must sink in much deeper than the larger ones. In traversing a field, the tracks made by the wheels are simply an extension of the depressions formed while standing still. In the case under consideration, both pairs of tracks will be 10in. wide, but those formed by the smaller wheels will be considerably deeper than those formed by the larger ones.

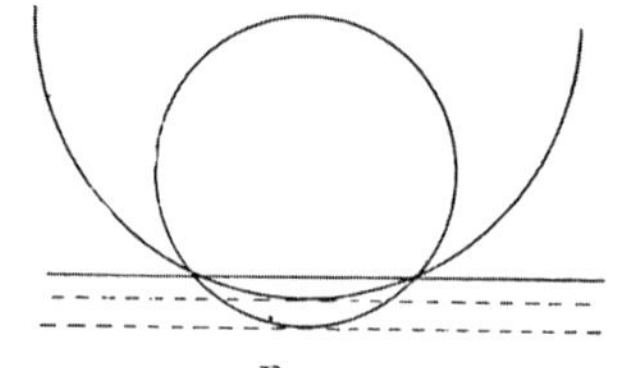

FIG. 1.
Diagram illustrating the advantages of large diameter driven wheels.

This proves that the soil under the smaller wheels is subjected to the greater pressure, and consequently suffers most from "packing." Therefore, it is quite evident that from the aspect of soil compression the greater the diameter of the driving wheels the better.

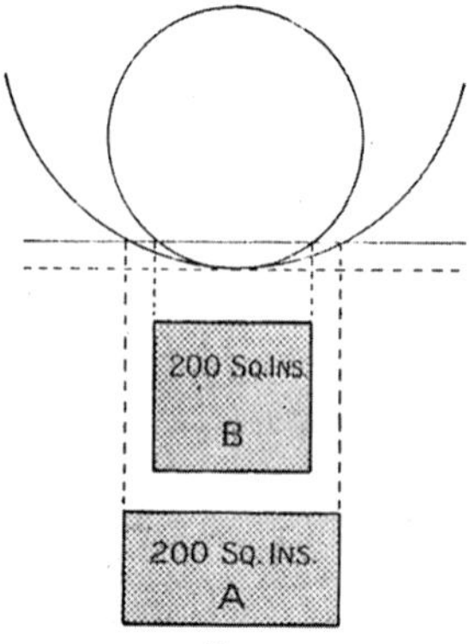

FIG. 2.
Diagram illustrating the area of frictional contact of driven wheels of different diameter and width of tread.

Width of Wheels.

If we increase the width of the smaller wheels, making them about 14in. wide, they will depend more on width than diameter for bearing surface, and will now find support when they have only sunk to the same depth as the large wheels. This will be understood by a glance at fig. 2. The solid horizontal line shows the ground level, and the broken line the depth to which the wheels sink. The shaded rectangles A and B show the

THE MOGUL TRACTOR.

THE 16 H.P. MOGUL TRACTOR.

ENGINE. One cylinder, 8×12in. (203×305 mm.)
POWER. 10 h.p. (2,000 lb., assuming 200 lb. pull per h.p.), 19 h.p. at belt pulley at 400 r.p.m.
FUEL. Paraffin.
WEIGHT. 50 cwt.
GEARS. One forward, one reverse.
CAPACITY. Three furrows; any belt machinery up to 20 h.p., including a 4ft. 6in. threshing outfit.
WHEELS. Four. Front, 3ft. × 6in. face; rear, 4ft. 6in. × 5in. face; extensions for wheels, 5in.
OVERALL DIMENSIONS. Length 11ft. 3in., width 4ft. 8in., height 5ft. 1in.

THE 25 H.P. MOGUL TRACTOR.

ENGINE. Two cylinders, 7×8in. (178×203 mm.)
POWER. 15 h.p. (3,000 lb., assuming 200 lb. pull per h.p.), 28 h.p. at belt pulley at 550 r.p.m.
FUEL. Paraffin.
WEIGHT. 90 cwt.
GEARS. Two forward, one reverse.
CAPACITY. Four furrows; any belt machinery up to 30 h.p.
WHEELS. Four. Front, 3ft. 4in. × 6in.; rear, 5ft. × 12in.; extensions for rear wheels, 5in.
OVERALL DIMENSIONS. Length 13ft. 6in., width 7ft. 6in., height 8ft. 4in.

shape and area of the depressions formed by a large and small wheel respectively.

It will be noticed that the area of the depressions formed is the same in both instances, viz., 200 sq. in., and that the wheels both sink in to the same depth.

In traversing a field, both pairs of wheels will now compress the soil to the same extent, but the 14in. wheels,

THE MOLINE TRACTOR.

ENGINE.	Two cylinders (horizontal), $4\frac{3}{4} \times 6$in. (120×152 mm.)	GEARS.	One forward and one reverse.
POWER.	6 h.p. (1,200 lb., assuming 200 lb. pull per h.p.), 12 h.p. at belt pulley.	CAPACITY.	Two or three-furrow plough or other work usually requiring five horses. A belt pulley is provided for stationary work.
FUEL.	Petrol.		
		WEIGHT.	25 cwt.

being a greater width than the others, will form tracks so much wider, and thus, while the degree of compression is the same in both cases, they will compress a greater area. This proves that what is lacking in diameter cannot be made up by width.

In deciding on the diameter of wheels, there is a limit beyond which it is not possible to go without incurring structural weakness, or adding to weight, as well as other considerations.

When the limit of diameter has been reached, the question naturally suggests itself, What is the most desirable width? We have found that, as width is added to the wheels, the area of land compressed increases, but that the degree of compression to which it is subjected decreases. The problem to be solved, therefore, is, Is it better to have a quantity of soil compressed slightly, or a small amount compressed considerably? This question is best answered when considering the influence of wheel width on adhesion and on draw-bar pull, but it can safely be said that of the two evils the former is the lesser. Therefore, when full advantage has been taken of diameter, under this heading wide wheels might be considered preferable to narrow ones.

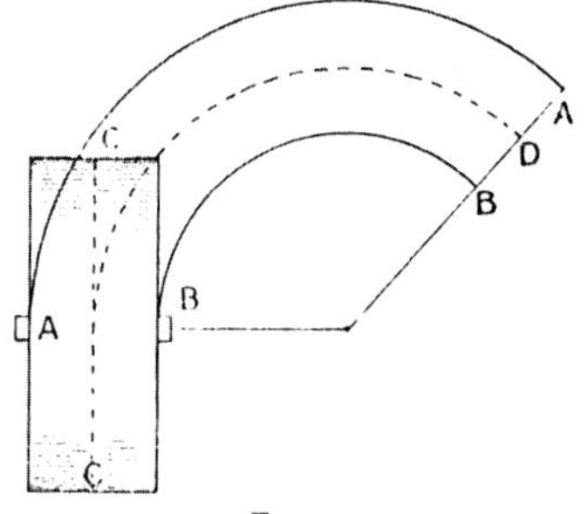

FIG. 3.
Diagram to illustrate the effect of width of wheel when turning.

Advantages and Disadvantages of Wide Wheels.

Like diameter, width must be kept within certain limits. The reader who has studied the explanation of the differential (see Chapter IV.) will best understand the difficulties attending the use of very wide wheels, for a very wide wheel is something like a tractor without a differential. Reference to fig. 3 will make it quite clear that in turning a corner the outside edge of a wheel must traverse a line A A, which is considerably longer than the line B B traversed by the inside edge. As the width of the wheel increases, the difference in length between the two tracks increases. There can only be true rolling contact between one line round the circumference of the wheel and the ground. If that point is the centre line C C and the dotted line C D, a certain amount of slipping must take place between the two outer edges and the ground. The inner edge must slip backwards, so to speak, and the outer edge be dragged forward.

If the wheel is fitted with gripping devices, the ones farthest from the centre, or in the case of cross strakes the ends of the cross strakes, will have to tear through the ground. This absorbs much power, and gives the wheel a tendency to keep straight on, which interferes with the steering, by causing the front wheels to side-slip when endeavouring to turn. On good firm ground 12in. to 14in. might be considered the greatest width permissible and all that would be desirable on a medium-powered machine, whilst makers have found that a diameter of 5ft. can be attained without incurring structural weakness or making the wheels unduly heavy.

On loose or soft ground a wide wheel is a great advantage, and fortunately on such ground the disadvantages of a wide wheel are not so much in evidence, as the inevitable slip between the two outer edges and the ground can more easily take place. The advantages of a wide wheel may be had, and the disadvantages avoided, by fitting to the driving wheels detachable extension rims for such work as ploughing on soft ground, preparing a seed bed, etc., and removing them when working on firm ground.

Influence of Wheels on Draw-bar Pull.

A good proportion of the power developed in the engine is absorbed in propelling a tractor, and, as stated in the early part of this chapter, the tractive effort necessary to propel the vehicle increases or decreases very greatly according to the class of surface over which it is travelling. If the surface is hard and smooth, the tractive effort necessary is slight. If, on the other hand, it is loose or compressible, the tractive effort necessary is proportionately great. In other words, whenever a lot of compressing is done by the wheels a lot of power is required to move them. A little reasoning will prove this. Power cannot disappear without leaving a corresponding amount of work done somewhere, and soil cannot be compressed without power is expended in compressing it; therefore, soil compression and power expended in causing it must always be in something like the same proportion to each other with a given width and diameter of wheel. It follows that, as the diameter of the driving wheels of a tractor is increased, the tractive effort necessary to propel the vehicle will be decreased, proportionately adding to the power available at the draw-bar.

Yet another point in favour of a large wheel is its ability to overcome obstacles. Anything but the very smoothest surface offers resistance to a wheel by its unevenness. The surface of a field is a good example. Every stone, clod, or tuft of grass is an obstruction that needs

THE MORSE LIGHT FARM TRACTOR.

ENGINE.	Four cylinders, 3½ × 5¼in. (89 × 133 mm.)	WHEELS.	Four. Rear, 3ft. 10in. diameter × 10in. face; front, 2ft. 6in. diameter × 5in. face.
POWER.	9 h.p. (1,800 lb., assuming 200 lb. pull per h.p.), 18 h.p. at belt pulley, 20 h.p. at 1,200 r.p.m.	CAPACITY.	Three-furrow 9in. plough, and any service an 1,800lb. draw-bar pull will perform.
FUEL.	Paraffin, starting on petrol.		
WEIGHT.	32 cwt.	OVERALL DIMENSIONS.	Length 10ft., width 4ft. 8in., wheelbase 7ft. 1in.
GEARS.	Variable forward and reverse by friction drive.		

power to overcome it, and it is a well-known fact that a large wheel will more easily overcome an obstruction than a small one. As an example, an American buggy with its 6ft. wheels will ride over a brick with ease, while a small stone may completely scotch the small wheels of a railway platform truck.

The number of obstructions to be overcome by a narrow wheel are not so numerous as those to be overcome by a wide one, and this explains the fact that, while a wide wheel does not compress the soil any more than a narrow one, it nevertheless requires more power to propel it.

The Chain Track.

Much of the foregoing chapter has been splendid argument in favour of the chain track, for in some respects the chain track may be compared to a wheel of infinite diameter.

The amount of track in contact with the ground can be extended to any convenient length; consequently, when the weight is distributed over such a large area soil compression is very slight, and the amount of power absorbed in propulsion very small. However, what is gained in one respect by a chain track is lost in another. Whilst a wheel has no moving parts to get out of order, a chain track has many. The track itself is built up of a great number of elements which are flexibly connected; besides, there are several guide wheels to keep the track in contact with the ground, and unfortunately all these moving parts work under the very worst conditions, being constantly smothered in grit or mud, which causes rapid wear and frequent breakage and clogging.

Wheel Grip or Adhesion.

Adhesion depends for its effectiveness on three conditions, namely: (1) The amount of wheel or chain track surface in contact with the ground; (2) the kind of gripping devices attached to the wheels or chain track; and (3) the weight of the vehicle.

As regards contact surface, with wheels this depends chiefly on the number of driving wheels employed and their width. A machine with two driving wheels will possess double the tractive grip of a machine of equal weight with only one driving wheel, other things being equal, and the width will be limited by the conditions referred to in an earlier part of the chapter: or, in the case of a machine running in the furrow, by the width of the furrow.

The contact surface of a chain track can be conveniently made much greater than that of a wheel, and where the chain track has failed (as it often has) to prove its superiority

THE NEVERSLIP TRACTOR. THREE SIZES—A, B, AND C.

A.

ENGINE.	Two cylinders, $6 \times 6\frac{1}{2}$in. (152 × 165 mm.)
POWER.	12 h.p. (2,400 lb., assuming 200 lb. pull per h.p.), 20 h.p. at belt pulley.
FUEL.	Petrol or paraffin.
WEIGHT.	50 cwt.
GEARS.	Two forward, one reverse.
CAPACITY.	Three 14in. furrows.
TRACKS.	Two, 5ft. × 12in.
OVERALL DIMENSIONS.	Length 9ft. 6in., width 5ft. 4in., height 6ft.

B.

ENGINE.	Four cylinders, $4\frac{3}{4} \times 7$in. (121 × 178 mm.)
POWER.	18 h.p. (3,600 lb., assuming 200 lb. pull per h.p.), 30 h.p. at belt pulley.
FUEL.	Petrol or paraffin.
WEIGHT.	54 cwt.
GEARS.	Two forward, one reverse.
CAPACITY.	Four 14in. furrows.
TRACKS.	Two, 5ft. × 12in.
OVERALL DIMENSIONS.	Length 10ft. 6in., width 6ft., height 6ft. 3in.

C.

ENGINE.	Four cylinders, $7 \times 8\frac{1}{2}$in. (178 × 216 mm.)
POWER.	40 h.p. (8,000 lb., assuming 200 lb. pull per h.p.), 60 h.p. at the belt pulley.
FUEL.	Petrol or paraffin.
WEIGHT.	100 cwt.
GEARS.	Three forward, one reverse.
CAPACITY.	Eight 14in. furrows.
TRACKS.	Two, 6ft × 1ft. 3in.
OVERALL DIMENSIONS.	Length 15ft., width 7ft. height 7ft. 6in.

in the matter of adhesion over a wheeled machine, it can be taken for granted that it has either not been fitted with adequate gripping devices, or the track has been too short or too narrow to cope with the work in hand.

Gripping Devices.

Of cross strakes, spuds, and spikes, cross strakes are the most effective and spikes the least. Cross strakes are usually made of angle iron set diagonally across the face of the wheel and reaching the full width, the only objection to them being that, fitted to wheels running on the unploughed land, they chop up the surface somewhat, causing the furrow slices to lie badly.

Spuds, especially those of wedge shape, grip very well, without unduly cutting up the surface. These, like cross strakes, are better set diagonally across the wheel, which arrangement makes them more or less self-cleaning, tends to prevent side-slip, and also gives a more continuous grip.

The depth of gripping devices on the majority of tractors imported into this country is usually about 2in. to 2½in., but it has been found that this depth can be advantageously increased to 3in. in bad weather, and the owner of a tractor would be well advised to provide his machine with two sets—one for when the land is wet and the other for when it is dry.

Weight.

Manufacturers appear quite unable to decide on what is the most desirable weight for a tractor: for, whilst some makers are turning out machines weighing between three and four tons, others are turning out machines with equally powerful engines weighing only 25 cwt. Weight, excepting as an aid to adhesion, is objectionable in every other respect. A heavy machine is expensive to build, expensive to maintain, difficult to handle, and has not the same capacity for work as a lighter machine of equal horse-power, as a greater proportion of the power is absorbed in its own propulsion. Moreover, a heavy machine is very damaging to compressible soils. In the present stage of development of the farm tractor, a personal opinion as to what is the most desirable weight would not be very authoritative, but on that point some very interesting views of the official observers at the

Scottish Tractor Trials, held by the Highland and Agricultural Society of Scotland in 1917, are recorded in their report.

Scottish Tractor Trials. Official Report on the Weight Question.

The report runs as follows: "With regard to weight, it will at once be apparent that, in many cases, the tractors shown have been evolved on the lines of the heavy steam

THE ONCE-OVER TILLER ATTACHMENT.

THIS CAN BE CONNECTED TO THE BELT POWER OF A TRACTOR OR DRIVEN BY A SMALL SEPARATE ENGINE. IT IS USED FOR PULVERISING THE SOIL AS IT LEAVES THE PLOUGH AND MIXING WITH IT FERTILISERS, SEED, ETC.

tractor. The most noteworthy feature of the present demonstration was the appearance of the light land tractor constructed on new lines. Several manufacturers appear to have departed from the idea that great weight is necessary for a tractor to do efficient work on the land. It was clearly shown at the demonstration that light machines, adequately provided with spuds, grip the ground and perform the work better than the heavier machines. Every drawback, such as slipping on soft land and inability to climb gradients, was aggravated by increased weight above a certain limit. Extra weight also increases the risk of breakage where stones are encountered and the danger of injury to the land through compression, this being very noticeable during the demonstration, especially when the driving wheel on the land travelled within a few inches of the previous furrow. Besides this, a heavy tractor is at a distinct disadvantage for the other and lighter forms of cultivation, such as grubbing, cultivating, seeding and harrowing, and also for harvesting, especially where the land is sown out with grass seeds.

"The light tractor is quite suitable, not only for ploughing, but for other farming operations, and therefore embraces all the usual requirements of a farm tractor, including the driving of a threshing mill and other farm machinery. It should be kept in view that for heavy stationary work proper anchorage must be provided. The only class of work for which the light tractor does not appear to be suited is road haulage. For road haulage weight is a necessity, as spuds cannot be used to give the necessary gripping power. This leads the committee to the conclusion that field work and road work are more or less incompatible, and that a light tractor, such as is suitable for land work, is not so useful for road haulage. In addition, for road work, a tractor should be sprung to minimise the vibration, while for field work springs are undesirable, and only entail extra weight. The committee have come to the conclusion that, to suit conditions in Scotland, an efficient land tractor need not exceed 30 cwt. in weight. In point of fact, several of the tractors which did excellent work at the demonstration were well within that weight."

CHAPTER VII.

DIFFERENT TYPES OF TRACTOR.

THE reader will gather from a study of this chapter that the farm tractor can never become standardised, as very many people appear to imagine, owing to the different qualities demanded to cope best with the conditions prevailing, not only in different parts of the country, but in the same locality and often in adjoining fields. Even for ploughing, as will be indicated, a tractor that is ideal for one class of land may not be at all suitable for another.

As no one type of tractor can ever be the ideal for every class of work, the owner who gets the best service from any given machine will be he who chooses the type that is best able to perform the job it will most often be called upon to do, and which will make the best compromise of all the other jobs.

To choose a tractor out of the dozens on the market which will best suit his own particular conditions, any farmer, without experience or expert advice to guide him, is up against a stiff proposition.

However, much of the fog is cleared away when it is realised that most of the minor differences in design have no influence on the work, and that many machines as wide apart in design as the two poles are identical as regards their qualities. As a matter of fact, of the great number of machines on the market, all differing from each other enough in detail to be confusing to the beginner, less than half a score can be chosen that possess differences in design which make them peculiarly suitable or otherwise for a particular class of work or kind of land, and the differences that matter ultimately resolve themselves into a question of horse-power, weight, width, and disposition of the wheels and gearing.

Something has already been said on all these points, but no excuse is necessary for enlarging on them when it is realised that success or failure depends so much on the choice of the right machine.

Different Arrangements of Driving Wheels.

When purchasing a tractor primarily intended for ploughing, or even when hiring a machine for that purpose, the disposition of the wheels should receive the most careful consideration. Many different arrangements of wheels are employed by different makers, each arrangement possessing certain merits, and each having its limitations. As a rule, the manufacturers and agents, instead of being content to claim for their tractors the particular merits which they possess, try to pass them off as the ideal machines for every class of work.

In view of the confusion that exists regarding the utility of different arrangements of wheels, the merits and demerits of the different arrangements will be somewhat fully dealt with.

No doubt, for any other operation than ploughing, the two-track, four-wheeled machines are the best, and under certain conditions they are also the best for ploughing, but under other conditions the best results can only be obtained by a modified arrangement of the wheels.

Four-track, Four-wheeled Machines.

Of course, in choosing a tractor for ploughing, as well as for any other operation, wheel grip is of very great importance. Obviously, the best grip will be obtained by the machine having two driving wheels of ample width—12in. to 14in. As the kind of ploughing favoured in this country is that from 8in. to 10in. wide, the wide-wheeled tractor must of necessity run with both its driving wheels on the unploughed land. Besides, many of the wide-wheeled tractors are four-track machines; that is, the front wheels are set closer together than the driving wheels, making it impossible to run in any width of furrow. While this is no drawback on some soils, and even advantageous on a few, on others the effect of the weight is very detrimental, especially to some clays, which, when not fairly soft, are too hard to plough at all. Besides the harm done to such soils due to "packing," they are all the stiffer to turn up after the passage of the tractor wheels, and every subsequent operation in the preparation of a seed bed is made more difficult than would otherwise be the case.

THE OVERTIME THRESHING. For Specification see Page 115.

Two-track, Four-wheeled Machines.

When it is considered that the advantages of the wide-wheeled tractor are outweighed by the disadvantages due to its weight all being on the unploughed land, the two-track, four-wheeled machine with two narrower driving wheels (usually 10in. wide) will be more suitable. With this, when using a 10in. plough, it is possible to run with the two off side wheels in the furrow, which has the effect of reducing the soil compression by half. Owing to the narrowness of the wheels some adhesion is sacrificed, but to some extent this is compensated for, as, the soil not

THE PETTER-MASKELL PLOUGH AND TRACTOR.

A Driver's Seat and Steering Gear are Provided.

Engine.	Two cylinders, 4¾ × 4½in. (121 × 114 mm.)	Capacity.	Two-furrow plough, cultivating, mowing, and reaping, driving farm machinery, electric lighting, pumping, etc.
Power.	7 h.p. (1,400 lb., assuming 200 lb. pull per h.p.), 12-14 h.p. at belt pulley.		
Fuel.	Kerosene, starting on petrol.	Wheels.	Two. Driving wheel, 3ft. diameter × 1ft. 3in. face; side wheel, 3ft. diameter × 3in. wide.
Weight.	24 cwt.		
Gears.	One forward, one reverse.		

Overall Dimensions. Length 10ft. 3in., width 5ft. 6in., height 4ft. 6in.

being compressed to such an extent, it will not require the same power to turn it up. Another advantage is that, as there is only one comparatively narrow wheel travelling on the unploughed land, the surface is not chopped up to such an extent by the spuds or strakes as is the case with the two wider wheels. This makes it possible to lay the work nicely without having to use a plough with unduly long breasts.

Chain-track and Two-track Three-wheeled Machines.

There is a great deal of land in this country on which weight of any kind is so harmful that many farmers who cultivate such land do not consider the tractor can ever supersede the horse for ploughing. On this kind of land the horses all walk in the furrow, so that no weight whatever is borne on the surface. One farmer in Bucks., who ploughs 300 acres of similar land, bought one of the best makes of tractor (but one that is not adapted to running in the furrow) four years ago, and, after trying it the first year, decided that it was so harmful to the soil that he has

THE "SIX-TWELVE" SAMSON SIEVE-GRIP.

ENGINE.	One cylinder, 7×9in. (178×229 mm.)	WEIGHT.	38 cwt. 3 qr. 14 lb. (shipping weight).
POWER.	6 h.p. (1,200 lb., assuming 200 lb. pul per h.p.), 12 b.h.p. at belt pulley at 450 r.p.m.	GEARS.	One forward, one reverse.
		WHEELS.	Three. Rear wheels, 3ft. 4in. × 1ft. 2in. wide; front wheel, 2ft. 4in. × 1ft. 3in.
FUEL.	Petrol or kerosene.		

OVERALL DIMENSIONS. Length 11ft. 6in., width 4ft. 6in., height 3ft. 10in.

since only used it for belt work. The above case is typical of the ignorance which prevails regarding the utility of different types of machine. The Bucks. farmer chose the worst possible type of machine for his particular kind of land when, had he known better, he might have purchased one that would have done his ploughing actually better than horses in every respect.

The correct tractor for the class of land alluded to is either the chain track, or else the two-track, three-wheeled machine possessing one large driving wheel and steering wheel running in the furrow, and an idler wheel, or small driven wheel, running on the unploughed land : the idler or small driven wheel only being to preserve the balance, and carrying little weight. With such a tractor practically all the weight is on the bottom of the furrow, the same as with a team of horses. In the case of the tractor, however, when using a three-furrow plough the wheels only traverse every third furrow, whereas with horses hauling a single plough they traverse every furrow, forming a continuous "pan."

A Popular Fallacy.

Three good horses would be about equal in weight to a medium-powered tractor. As the horses would have to

THE "TEN-TWENTY-FIVE" SAMSON SIEVE-GRIP.

ENGINE.	Four cylinders, $4\frac{1}{4} \times 6\frac{3}{4}$in. (108 × 171 mm.)	GEARS.	One forward, one reverse.
POWER.	10 h.p. (2,000 lb., assuming 200 lb. pull per h.p.), 25 b.h.p. at belt pulley at 650 r.p.m.	WHEELS.	Three. Rear wheels, 3ft. 4in. diameter × 1ft. 6in. wide, 6in. road band; front wheel, 2ft. 4in. diameter × 1ft. 3in. wide.
FUEL.	Petrol or kerosene.	OVERALL DIMENSIONS.	Length 12ft. 1in., width 5ft. 2in., height 4ft. 2in.
WEIGHT.	46 cwt. 2 qr. (shipping weight).		

travel three times over the ground to do as much work as a tractor travelling once, it will at once be apparent to the reader that it is only a popular fallacy that a tractor causes more soil compression than a team of horses doing an equal amount of work. Only having one large driving wheel, limited in width to the width of the furrow, this type of tractor does not possess the same adhesive qualities as one with two large driving wheels; nevertheless, it is often possible to work with it when the land is too wet for any other tractor, for, when the bottom of the furrow is dry, the condition of the surface does not matter, as practically all the driving is done by the large wheel.

Tractors which run with all their wheels out of the furrow are rather difficult to steer, and to aid steering some makers supply an automatic steering device, which consists of a small furrow wheel attached to the off side front wheel.

THE SAUNDERSON TRACTOR.

ENGINE.	Two cylinders, $5\frac{1}{2} \times 8$in. (140 × 203 mm.)	WEIGHT.	52 cwt.
		GEARS.	Three forward, one reverse.
POWER.	17 h.p. (3,500 lb., assuming 200 lb. pull per h.p.), 22-24 h.p. at belt pulley.	CAPACITY.	Four furrows, threshing, hauling, etc.
FUEL.	Paraffin.	WHEELS.	Four. Diameter 4ft. × 10in. face.

OVERALL DIMENSIONS. Length 12ft., width 5ft. 6in., height 7ft. 6in.

Those machines which run with their off side wheels in the furrow are more or less self-steering, and it is quite possible for the driver to dismount and walk by the side of his charge along a straight furrow.

As a self-lifting plough requires a certain amount of attention on the part of the driver, a tractor which runs in the furrow is more adapted to the hauling of such a plough. Some arrangements of wheels which possess no outstanding qualities are determined by structural considerations; for instance, a machine with only one driving wheel requires no differential and less driving gear, therefore it can be made more cheaply.

The Various Types of Engine.

There are single, double, and four-cylinder vertical engines, and single and double-cylinder horizontal engines, the two kinds most in evidence being the two-cylinder horizontal engine developing its maximum power at about 500 r.p.m., and the four-cylinder vertical engine developing its full power at about 1,000 r.p.m., popularity being about equally divided between the two types.

The makers of the four-cylinder vertical engines claim that a power plant can be built on these lines which is much lighter, freer from vibration, and more efficient than a two-cylinder horizontal engine of equal power; whilst the makers of the two-cylinder horizontal engine claim that their design is more adapted to tractor use because, owing to the slow speed, less power is absorbed by the reducing gear, and because it can be conveniently laid across the frame with the crankshaft at right angles to the direction of travel, thus dispensing with right-angle drives employing bevel or worm gearing. They also claim for it simplicity owing to its fewer working parts, and longer life on account of its slow speed.

As a matter of fact, as far as present experience can show there is little to choose between the two types. Each possesses qualities which the other lacks. Some users swear by one and some by the other, but it is probable that when adhesion has been increased to its maximum by every means possible excepting by weight, engines with four or more cylinders will become universal owing to their comparative lightness.

Seven Distinct Types of Tractor.

It was stated in the early part of this chapter that the great variety of tractors on the market do not differ so much in character as they do in appearance, and that the number of machines which in principle of design possess fundamental differences which are an important consideration when making a choice are very limited. In the course of the chapter, and in other parts of the book, it has been indicated that different soils and different classes of work call for certain variations in the design of a tractor that cannot be embodied in any one machine.

To illustrate these points further, seven tractors, representing as many distinct types, are shown on the following pages. A brief specification of each type is given, and the outstanding features which distinguish one from the other are picked out and discussed more fully.

Each type is suitable for some particular class of work or kind of land, and every class of work or kind of land can be done well by one of these types, and though every tractor could not be strictly classified in the seven groups represented by the seven tractors referred to, a study of this chapter and a close examination of the forty or more examples illustrated in this book will show that they all conform in a greater or less degree to one or other of the seven types chosen. It must be understood that in most cases there are several manufacturers of similar types to the seven discussed.

Queries and Replies.

Readers who wish for further information regarding any particular tractor, plough, etc., or who may be in any difficulty regarding the running and maintenance of a tractor, and desire individual advice on any subject dealing with tractors and their work which is not contained in the book, are invited to communicate with the Editor of *The Agricultural Gazette*, who will be pleased to answer their queries through the post. All letters requesting information should be written on one side of the paper only, and should be addressed, "The Editor, *The Agricultural Gazette*, 20, Tudor Street, London, E.C.4," and a stamped addressed envelope for reply should be enclosed.

Alldays General Purpose Tractor.

A tractor intended to haul heavy loads along the highway must be well sprung to comply with the law and to minimise the shocks occasioned by the hard roads, and be fairly weighty to give it the necessary tractive grip on a smooth surface, especially on gradients; while for land operations it is becoming more evident every day that excessive weight is not conducive either to good or economical work, and that springs, if not objectionable, are at least superfluous.

In spite of the apparent incompatibility of the two classes of work, the Alldays General Purpose Tractor combines with a great measure of success the qualities demanded for both classes of work.

The gross weight of the machine is 55 cwt., and to give it the increased weight necessary for road haulage a ballast tank, capable of holding 1,200 lb. of water, is carried over the rear axle, giving it a capacity for dealing with a four-ton trailer load. Both front and rear axles are suitably sprung, and the rear springs are so arranged that it is possible to lock them out of action, thereby converting the rear suspension into a rigid chassis when required for land operation.

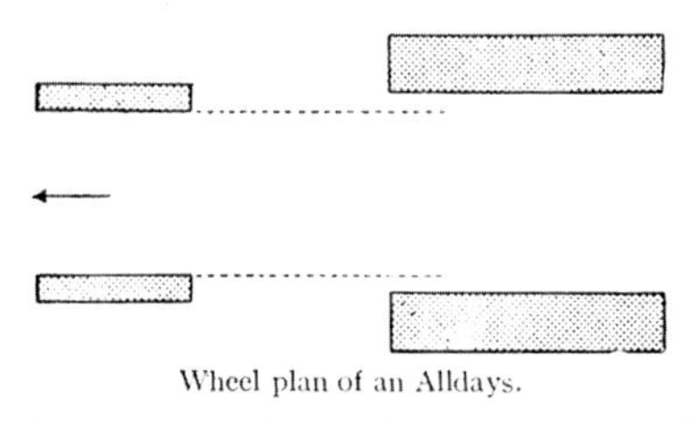

Wheel plan of an Alldays.

As will be seen by the wheel plan, the Alldays tractor is not intended to track in the furrow. By many 55 cwt. will be considered a heavy weight to be borne all on the surface, but those who consider their soil will stand it, and who want their machine to combine the qualities of a land tractor with those of a road tractor, will find the Alldays the nearest approach to their requirements. The powerful engine and suitably-placed belt pulley, geared down to 250 r.p.m., makes this an ideal plant for all kinds of belt work.

The Fordson Tractor.

If sufficient tractive grip can be obtained by a powerful tractor weighing less than 25 cwt., then the 22 cwt. Fordson represents a type which is approaching the ideal: for, be it remembered, weight is not necessary for any other purpose but to aid adhesion and give strength. Strength, however, can be had without exceeding a ton, by the combination of good material and good workmanship; therefore weight need only be considered from the point of view of adhesion. According to the report on the demonstration of tractors and ploughs, held by the Highland and Agricultural Society of Scotland (see Chapter V.), the official observers were of opinion that a tractor need not exceed 30 cwt. for ploughing

THE ALLDAYS GENERAL PURPOSE TRACTOR.

ENGINE. Four cylinders.
POWER. 30 h.p., draw-bar pull 2,000 lb.
FUEL. Paraffin, starting on petrol.

GEARS. Three forward, one reverse.
WEIGHT. 55 cwt.
CAPACITY. Three-furrow plough, road haulage, threshing, and general purpose work.

and ordinary cultivating operations, and the gradual trend of design in the direction of lightness certainly appears to uphold that view.

It is worth noting that the weight of the Fordson tractor is only 1 cwt. per horse-power, compared to the 3 or 4 cwt. per horse-power of some of the heavier machines on the market. There must be a time near at hand when weight in relation to horse-power will be better understood, and when a certain weight per horse-power will be known to be the best for a given class of work. Weight per horse-power will then be as important a term in tractor parlance as it is in aeroplane engine construction at the present day.

In addition to anticipating the probable future developments in the matter of weight, the makers of the Fordson appear to have got ahead of rival firms in general design in a manner worthy of our attention. In seeking to improve a complicated piece of machinery, a good engineer endeavours to dispense with some part of it, whilst a bad one adds something to it. The former principle, which has apparently been adhered to very closely by the designers of the Fordson, has resulted in the production of an efficient little tractor that is the embodiment of simplicity: and simplicity must have a greater influence in making the tractor popular than any other factor—first, because it makes possible low cost of production and low selling price, and thus appeals to the farmers as a profitable investment; and, secondly, because the average farmer has not yet acquired that confidence in machinery possessed by the manufacturer and the average townsman, consequently the simpler a machine is the more readily he will take to it.

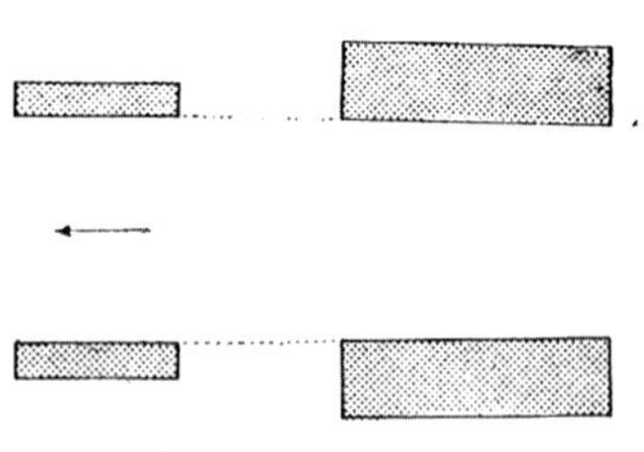

Wheel plan of a Fordson.

A glance at the Fordson tractor reveals in a striking manner how heavy and costly impedimenta can often be dispensed with without impairing efficiency, the most noteworthy example being the entire absence of the usual frame, with its cross members and brackets. The machine is merely a power unit and transmission gear *en bloc* mounted on two pairs of wheels, together with the simplest of control and steering arrangements.

The whole of the transmission gear from the engine to the axle driving-shafts is enclosed and runs in oil, and yet there are only sixteen places in all dependent on the driver for lubrication; all the other parts are automatically lubricated from the crank chamber or the transmission case.

Other interesting features of the Fordson are the distinctive ignition system (see "Ignition," Chapter III.)

THE FORDSON TRACTOR.

THIS IS THE ORIGINAL FORD DESIGN PRESENTED TO THE BRITISH GOVERNMENT AND TAKEN OVER BY THE MINISTRY OF MUNITIONS AGRICULTURAL MACHINERY DEPARTMENT.

ENGINE.	Four cylinders, 4×5in. (102 × 127 mm.)	CAPACITY.	Three-furrow plough, general farmwork.
POWER.	Actual pull: Max. 3,000 lb., min. 2,000 lb., 25 h.p. at belt pulley at 1,000 r.p.m.	WHEELS.	Four. Driving wheels, 3ft. 6in. diameter × 12in. face.
FUEL.	Paraffin, starting on petrol.	OVERALL DIMENSIONS.	Length 8ft. 5in., width 5ft. 2in., height 4ft. 6in.
GEARS.	Three forward, one reverse.		
WEIGHT.	22 cwt.		

and the final worm drive to the rear axle (see "Transmission and Steering," Chapter IV.).

The Burford-Cleveland Tractor.

There are many types of chain track tractors. Some have one track and two wheels, some have double tracks and one wheel, while others like the Burford have two tracks only. They are all more or less alike in principle, and all have the same advantages and drawbacks, therefore one example will serve our purpose here.

The champions of the chain track claim that it does not compress the soil to the same extent as wheels; that the motive power absorbed in propulsion is not so great; and that a chain track is superior in gripping power to a wheel. That the chain track is very much superior to wheels as regards soil compression, a little reasoning will soon show.

It is not the weight that passes over a given surface, but the pressure exerted per square inch on that surface, which decides the degree of compression caused. The Burford, for instance, has a track area in contact with the ground of 600 square inches, and its total weight is about 3,000 lb.; therefore the pressure per square inch exerted on the soil would be 5 lb. A pair of fair-sized wheels the same width as the track alluded to, and carrying the same weight, could not be considered to have more than about 150 square inches of surface in contact with the ground under average conditions, which means that they would exert a pressure of 20 lb. per square inch on the soil.

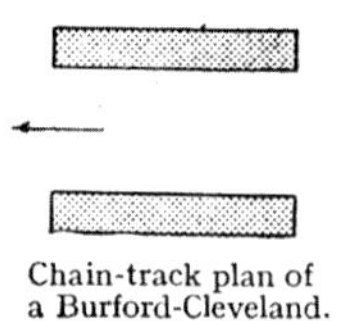
Chain-track plan of a Burford-Cleveland.

The motive power required to propel a vehicle is more in proportion to the amount of compression it causes than to its weight, as explained under the heading "Influence of Wheels on Draw-bar Pull" (Chapter VI.); therefore the chain track tractor would be expected to absorb much less power in propulsion than a wheeled machine of similar weight, and the sponsors of the chain track loudly proclaim that to be the case. The advantages gained in this respect, however, are not so great as they would at first appear, for a goodly amount of power is expended in turning the track and its various guide wheels among the conditions of grit and dirt which nearly always prevail.

As regards gripping power, makers and agents are in the habit of making a comparison between the respective gripping areas of a chain track and a wheel, and many people have been misled into the belief that a chain track is as much superior to a wheel in gripping power as it exceeds it in contact area. The fact is, the weight is so distributed over a chain track that it only grips the ground

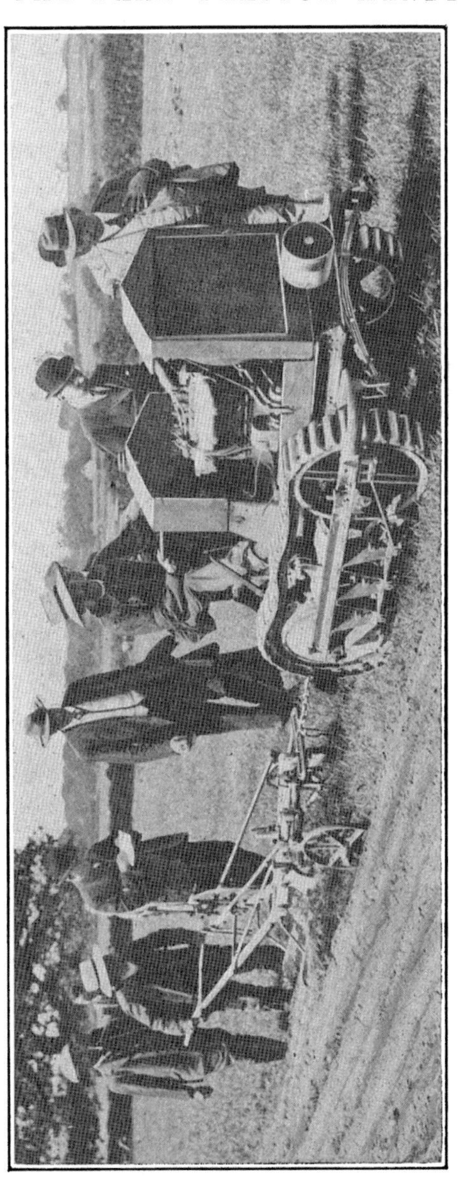

THE BURFORD-CLEVELAND TRACTOR.

HAULING A TWO-FURROW PLOUGH. THIS MACHINE IS OF THE CHAIN TRACK VARIETY.

ENGINE.	Four cylinders, $3\frac{3}{8} \times 5\frac{1}{8}$in. ($85 \times 130$ mm.)	GEARS.	One forward, one reverse.
FUEL.	Paraffin, petrol for starting.	WEIGHT.	27 cwt.

lightly, whereas a wheel sinks in to some extent and forms itself a hard bed on which to get a grip. Nevertheless, if a chain track has a fair contact area, and is fitted with adequate gripping devices, it will work under conditions that would be impossible for a wheel of ordinary diameter and width.

The one objection to a chain track is the great number of moving parts of which it is formed, and the very bad conditions of grit and dirt under which all these moving parts must work.

To return from the general to the particular, it will be noticed that a rather small diameter belt pulley is rigidly fixed to the crankshaft of the Burford and placed in front of the radiator at right angles to the main frame. This arrangement, which is very inconvenient, does not enable the belt to be stopped without stopping the engine unless a countershaft provided with fast and loose pulleys is employed, and the angle at which the pulley is set would entail a certain amount of manœuvring to get the machine in position. The absence of a governor is somewhat detrimental to the capabilities of the machine for belt work.

The Overtime.

As regards its arrangement and width of wheels especially, this machine is typical of a tractor which should eventually become very popular in this country owing to its adaptability to various conditions.

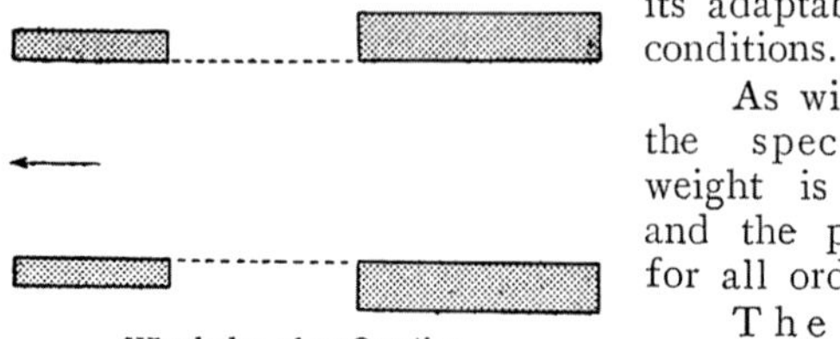

Wheel plan of an Overtime.

As will be seen from the specification, the weight is not excessive, and the power is ample for all ordinary work.

The engine is governed, and has a belt pulley placed in a convenient position and under the control of the main clutch.

The wheel plan shows the steering and driving wheels to be in line, the driving wheels being 10in. wide. This means that the two off side wheels can be run in the furrow, at the same time allowing a 10in. plough to be used. With half the weight in the furrow, only 21½ cwt. is borne by the unploughed land, against three times that weight when using some of the heavier type machines which do not run in the furrow. The wheels being of limited width, they do not obtain the same grip as those of a wider pattern, and obviously would not be so good for work on a slippery surface, or on loose ground such as a partly prepared seed bed, where width is a great advantage. However, this

THE OVERTIME TRACTOR.

ENGINE.	Two cylinders, 6×7in. (152 × 178 mm.)	GEARS.	One forward and one reverse.
POWER.	12 h.p. (2,400 lb., assuming 200 lb. pull per h.p.), 24 h.p. at belt pulley.	CAPACITY.	Four furrows, full size thresher, etc.
FUEL.	Paraffin.	WHEELS.	Four. Driving wheels, 4ft. 4in. × 10in. face.
WEIGHT.	39 cwt.	OVERALL DIMENSIONS.	Length 11ft. 10in. × 6ft. 1in.

difficulty can be quite overcome and certain advantages gained by the judicious use of a pair of extension rims fitted with extra cleats.

A pair of 6in. extension rims would increase the width of the wheels to 16in., and make them very suitable for cross ploughing, cultivating, harrowing, etc.; besides, when conditions are too wet for the wheels to travel in the furrow when ploughing, the addition of a pair of extension rims would often make it possible to continue work, running with all the wheels on the unploughed land. Extension rims will also check the tendency to side-slip which is caused, when running out of the furrow, by the oblique set of the draught rope and the narrowness of the wheels.

Again, when the bottom of the furrow is dry, but the surface is too wet to carry a 10in. wheel, matters can be greatly improved by using one extension rim only, fitted to the wheel which travels on the surface. It will be clear, therefore, that the arrangement of wheels employed on machines of this type is the best for all-round land work, and it can safely be said that a machine built on these lines, but a little lighter, and fitted with a two-speed gear, would be the most suitable for the average farmer who requires a tractor capable of tackling many different classes of work under various conditions of weather and soil.

The Whiting-Bull.

The weight of the Whiting-Bull is moderate, and, as the accompanying wheel plan shows, is so distributed as to be mostly on the one large driving wheel and the steering wheel, the land wheel just carrying sufficient to give the machine balance and stability. As the large driving wheel and steering wheel both run in the furrow, very little weight is borne by the unploughed land; therefore this type of tractor is just the right kind to plough those heavy compressible soils on which weight of any kind is so objectionable: and excepting the chain track type, it is the only type of tractor that can ever supplant the horse for ploughing in many parts of England.

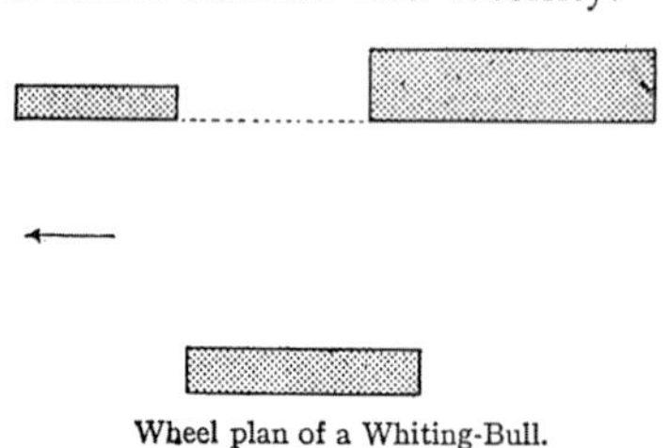

Wheel plan of a Whiting-Bull.

To obtain sufficient grip with one large driving wheel (the small land wheel, of course, not having much driving power), it is necessary to make the tread fairly wide, which means that very narrow ploughing cannot be indulged in. However, it will soon be realised that with the larger, stronger tractor-drawn implements now coming on the market, wide ploughing is not so difficult to get down to a seed bed as has hitherto been the case with horses.

Owing to the limited grip obtained by the arrangement of wheels, machines of this type are not well adapted for road haulage, nor so good as a machine with two large

THE WHITING-BULL TRACTOR.

An American Machine, with Single Front Steering Wheel in Line with the Driving Wheel; a Land Wheel opposite the Driving Wheel runs loosely on its Axle.

Engine. Two cylinders, $5\frac{1}{2} \times 7$in. (140 × 178 mm.)

Power. 12 h.p. (2,400 lb. assuming 200 lb. pull per h.p.), 24 h.p. at belt pulley.

Fuel. Paraffin.

Weight. 44 cwt.

Gears. One forward, one reverse.

Capacity. Three furrows, light land four furrows, threshing, chaff cutting, binding, reaping, harrowing, potato digging, planting, cutting, stacking, etc.

Wheels. Three. Front wheel 30 × 6in. centre rib 3in., land wheel 40 × 8in., main drive wheel 60 × 12 or 14in.

Overall Dimensions. Length 13ft. 7in., width 6ft. 6in., height 6ft. 6in.

driving wheels for other cultivating operations than ploughing.

The engine of the Whiting-Bull is a departure from usual tractor practice. It will be observed from the specification that a two-cylinder horizontal engine with a bore and stroke of $5\frac{1}{2}$in. × 7in. gives 24 h.p.—a slightly higher power than is usually obtained from a two-cylinder engine of similar dimensions. This, it is claimed, is obtained by opposing the cylinders—that is, placing them at opposite sides of the crankshaft instead of side by side. By the arrangement better balance of the moving parts is secured, which permits lighter construction and results in a higher engine speed—in this case, 750 r.p.m. instead of the usual 500.

Wyles Motor Plough.

The self-contained motor and plough—a type which originated in this country—is fast gaining favour in many parts of the world, and when it is considered that a machine like the Wyles can be bought for a price that does not much exceed the price of a good horse, this is not to be wondered at. Besides, the principle of design is sound, and it is quite within the range of possibility that the single unit idea will be adopted more in the construction of powerful machines than it has been hitherto. There are already several machines of this class on the market, ranging from 10 h.p. to 20 h.p.

As practically all the weight of both tractor and plough is carried by the driving wheels, the total weight of the combination can be kept very low without interfering with the tractive grip, with the result that it is the most economical ploughing implement on the market.

Wheel plan of a Wyles plough.

The Wyles is a good example of several small low-powered types that have made their appearance since the tractor boom started. It is a British-made machine, and has several good qualities which should particularly commend it to the small farmer.

It is capable of pulling a two-furrow plough through medium soil at $2\frac{1}{4}$ m.p.h., while its lower gear of $1\frac{3}{4}$ m.p.h. enables it to tackle heavier soils with the same plough.

The plough can be detached and any other cultivating implement hung on behind in its place; while for mowing, reaping, and hauling the makers supply a special attachment, comprising a third wheel and a steering arrangement.

A useful belt pulley is fitted, which can be controlled by the clutch. The engine is governed, and the entire transmission gear runs in oil. No differential gear is fitted, but one wheel can be thrown out of action when turning, which permits of very short turning on a narrow headland.

THE WYLES MOTOR PLOUGH.

The Lower View shows the Machine converted to a Tractor and hauling a Mower.

Engine.	Single-cylinder, 5×6in. (127 × 152 mm.)	Fuel.	Petrol or Wyles fuel oil.
		Gears.	Two forward.

Capacity. Two-furrow plough; belt pulley provided for light stationary work; an attachment can be supplied to convert the machine to a tract or for hauling a mower or self-binder.

Weight. 21 cwt., including two-furrow plough.

With the Wyles the driver walks behind in the same manner as with a horse plough. In one or two other machines a seat is also provided, and walking or riding is optional.

The Omnitractor.

The weight of this type of tractor places it in the heavy class, and being a four-track machine with wide wheels,

THE OMNITRACTOR.

ENGINE.	Two cylinders, 6½ × 9in. (165 × 229 mm.)	CAPACITY.	Four furrows 9-12in. wide and 6-9in. deep., according to land.
POWER.	20 h.p. (4,000 lb., assuming 200 lb. pull per h.p.)	WHEELS.	Four. Driving wheels, 5ft. diameter × 1ft. 4in. face; front wheels, 3ft. 6in. diameter × 9in. face.
FUEL.	Paraffin.		
WEIGHT.	66 cwt.		
GEARS.	Two forward, one reverse.		
OVERALL DIMENSIONS.	Length 11ft., width 7ft.		

it is not adapted to run with two of its wheels in the furrow, therefore all the weight must be borne by the unploughed land.

Its substantial weight and wide wheels give it exceptional gripping power on firm or light land where soil compression is not a serious consideration. On such land, however, adhesion is easier to obtain, and it is a question whether quicker and more economical work could not be done by a lighter tractor of equal power.

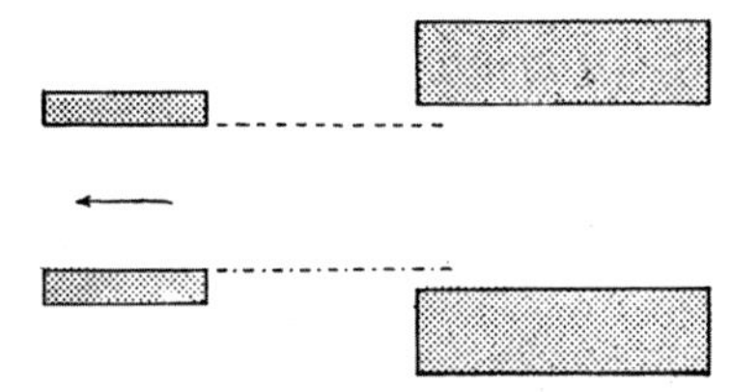

Wheel plan of an Omnitractor.

The only operation calling for exceptional weight is road haulage, and this three-speed machine, being well sprung and having substantial wheels fitted with regulation strakes, is well equipped for that class of work. The powerful governed engine and suitably placed belt pulley also fit it for all classes of belt work.

THE GARNER ALL-PURPOSE TRACTOR.

ENGINE.	Four cylinders, $4\frac{1}{4} \times 5\frac{1}{2}$in. ($108 \times 140$ mm.)
POWER.	29 h.p. on belt.
FUEL.	Paraffin, starting on petrol.
WEIGHT.	34 cwt.
WHEELS	Four: two rear, 3ft. 4in. diam., 10in. face; two front, 2ft. 6in. diam., 4in. face.
GEARS	Three forward, one reverse.
SPEEDS	First and reverse $1\frac{1}{2}$ m.p.h., second $2\frac{3}{4}$ m.p.h., third 5 m.p.h.
CAPACITY	Three furrows on all ordinary land, two on stiff land; disc harrowing and cultivating; threshing and general belt work.
OVERALL DIMENSIONS.	Length 10ft. 6in., width 5ft. 6in., height 4ft. $7\frac{1}{2}$in.

CHAPTER VIII.

PLOUGHS.

THESE may be divided into three classes: (1.) The independent plough, hauled by a rope or chain, and requiring an attendant to manipulate the levers for setting in, lifting out, regulating the depth, and steering.

(2.) The power-lift plough, which is under the control of the tractor driver, and the source of power for lifting is the engine, which acts through a clutch on one of the plough wheels attached to a cranked axle, or lifts the plough bodily by a crane-like attachment.

(3.) The self-contained tractor and plough, both arranged on the same frame, which is more or less balanced on a pair of driving wheels.

The great advantage of the power-lift plough over the independent plough lies in the fact that it can be operated by the driver of the tractor, thus cutting the wage bill by half. On even ground there is nothing to choose between the work of a power-lift plough and that of an independent plough. However, on uneven ground, and where soft places are encountered, the independent plough has the advantage, because the necessity for frequently adjusting the depth levers requires the undivided attention of the ploughman.

The power-lift plough has no actual steering mechanism, but is semi-rigidly coupled to the tractor and automatically follows in its track; therefore it is more adapted for use behind a machine which runs in the furrow, or one that is provided with an automatic steering device, which arrangements give the driver more freedom to attend to the plough, besides ensuring straighter work.

The self-contained tractor and plough, being as much a tractor as a plough, is dealt with in Chapter VII.

Influence of the Tractor on Ploughing.

It is often said that the quality of the ploughing depends entirely on the plough and the ploughman, and that the tractor is simply the motive power which takes the place

SAUNDERSON TRACTORS. SPECIFICATION ON PAGE 105.

of the horse. This is not strictly true, the tractor having a great influence on the quality of the ploughing, as will be gathered from the foregoing sections dealing with gripping devices, width and disposition of tractor wheels, and weight and horse-power. A lot also depends on the driver of the tractor; the best combination of plough, ploughman, and tractor has no chance of turning out good work if the driver is inexperienced or careless, for to a great extent the straightness of the work depends on the driver, especially if the tractor is one which runs with all wheels out of the furrow; and, again, the responsibility for keeping the ploughing square at the ends of the furrow rests partly on the driver, for if he runs too wide or too near, the plough will either miss the end of the furrow altogether or go in with a curve.

There are, no doubt, good ploughs and bad ploughs, but as a rule the majority of ploughs are suitable for some particular class of work or kind of land, and their suitability or otherwise is determined chiefly by the design of breasts employed and the pitch of the bottoms—that is, the width of furrow it cuts. The best of ploughs on the most suitable land, however, will not do good work if the breasts, coulters, skims, etc., are not correctly set.

Depth and Width of Furrows.

As a rule, the correct depth to plough is as deep as possible without bringing up the subsoil, and the correct width is that which sets the slices up at a sharp angle. When the furrow slices are set up at the correct angle, width and depth will be in about the same proportion to each other; therefore, as soil depth varies very considerably, ploughs should be adjustable as regards width. Unfortunately, this does not seem to be understood by many plough makers and importers, for a great number of set ploughs are on the market, many of which are so wide as to be only suitable for very deep soils or for cross ploughing.

A 14in. furrow plough working only 4in. deep lays the work too flat, whilst a 10in. furrow plough working 8in. deep lays it too much on end. Fig. 1 shows, in section, ploughing of a set width (10in.) done to various depths. A shows that 6in. × 10in. is about right, B that 8in. × 10in. sets it too much on edge, whilst C shows that 4in. × 10in. lays it too flat.

Weight of Plough.

It is sometimes argued in favour of the set plough that, weight for weight, it is more rigid than the adjustable plough. Even if this is so, 2 or 3 cwt. more or less in the weight of a plough is immaterial.

The draw-bar pull required to haul a 7½ cwt. plough, apart from the cutting and turning of the soil, is not great, and might be put at anything between 60 and 120 lb.—say 90 lb. A 10 cwt. plough on that basis would require a draw-bar pull of 120 lb.—an increase of 30 lb. over the lighter plough. Considering that the cutting and turning of three furrows on fairly stiff land involves a draw-bar pull of, say, 3,000 lb., it is quite clear that the extra 30 lb. pull demanded by the heavier plough is quite insignificant.

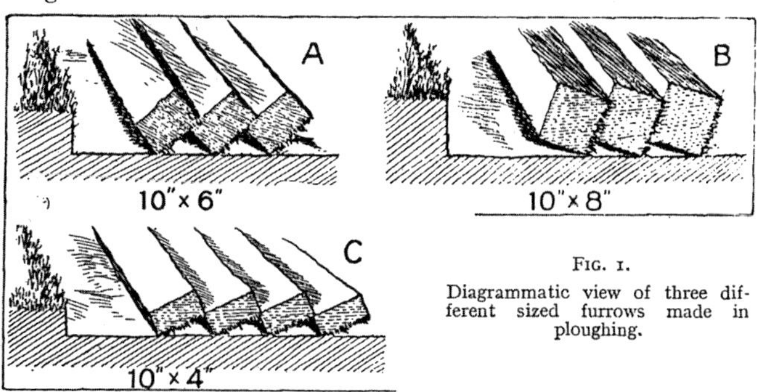

FIG. 1.
Diagrammatic view of three different sized furrows made in ploughing.

In face of the above, it is quite a popular misconception that a heavy plough is objectionable on the ground that it appreciably increases the draught. As a matter of fact a heavy plough may be of light draught and a light plough may be of heavy draught, the difference depending on the type of breast fitted. Moreover, the draught of either is affected by the setting of the breasts, the setting of the coulters, and the position of the draught rope, chain, or rigid coupling (we will call it the draught rope).

Breasts.

Fig. 2 shows four kinds of breast. A is the short comparatively straight American or Continental type; B is the long sweeping British type; C is similar to B, but has its rear under edge cut away, giving it some of the characteristics of A; whilst D is a compromise between the three.

The short type breast A is admirable for cross ploughing, and suitable for first ploughing on loose, light land. The breasts being short and straight, they do not lay over the furrow slices on stiff land, but leave them stuck up on end, with a tendency to fall back into the furrow. Owing to the fact that the frictional surfaces are small, and that the soil is not displaced to such an extent as with a longer breast, the draught of a plough fitted with this type of breast is comparatively light.

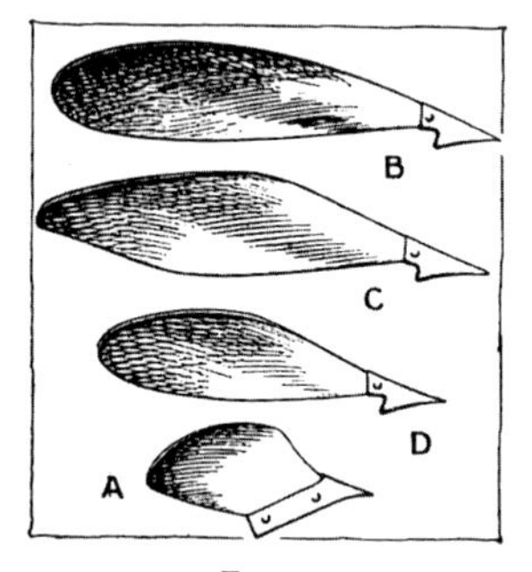

FIG. 2.
FOUR TYPES OF PLOUGH BREAST USED FOR WORK ON VARIOUS KINDS OF LAND.
A. American or Continental pattern.
B C. Long British types.
D. An intermediate pattern.

Really stiff soil can only be laid well by a long sweeping breast with a good curl in it, and when it is necessary to have the furrow slices pressed closely together to bury the grass, or for sowing broadcast, the types B or C are the correct ones to use. C is an improvement on B; having its under edge cut away at the rear a lot of frictional surface is removed, whilst the pressure on the furrow slice is just where it is required most, viz., on its upper edge.

A lot of the most successful Government ploughing has been done by ploughs fitted with the type of breast shown at D. These breasts are rather shorter than those shown at B or C, and have not such a pronounced curl; consequently the draught of a plough so equipped is lighter than one fitted with the breasts B or C, but heavier than one fitted with the short type shown at A. For anything but the very stiffest soil these breasts are to be recommended. Of course, if the land is to be fallowed or will subsequently be cross ploughed, a badly-laid furrow is more or less immaterial; therefore, in the interest of economy, it is wise to use the plough which is the easiest to pull.

THE 20 H.P. TITAN TRACTOR.

ENGINE. Two cylinders, 6½ × 8in. (165 × 203 mm.)
POWER. 12 h.p. (2,400 lb., assuming 200 lb. pull per h.p.), 22 h.p. at belt pulley at 500 r.p.m.
FUEL. Paraffin.
WEIGHT. 55 cwt.
GEARS. Two forward, one reverse.
CAPACITY. Three or four furrows; any belt machinery up to 22 h.p., including 4ft. 6in. threshing outfit.
WHEELS. Four. Front, 3ft. × 6in. face; rear, 4ft. 6in. × 10in. face; extensions for rear wheels, 5in.
OVERALL DIMENSIONS. Length 12ft. 3in., width 5ft., height 5ft. 6in.

The ideal plough is one with interchangeable breasts of various kinds, and a few makers are beginning to make such ploughs.

Adjusting the Breasts.

Most breasts are held by adjustable stays, whereby the rear ends may be pressed outwards or drawn inwards by adjusting the screwed stay. To lay the work well it is necessary to have them set well out, but it should be remembered that as they are set out the draught is increased, therefore in the interest of economy they should not be set out more than is necessary.

Disc Coulters.

Although disc coulters have been on the market for so many years, it is remarkable that a great number of farmers had never used them or even seen them in use until the advent of the Government plough. They are a great improvement on the ordinary knife coulters, as they cut cleanly through the toughest turf or through any kind of tough tangled growth such as couch grass or water grass, which has a way of clinging to knife coulters and "choking" the plough. It may be that many farmers have avoided disc coulters on account of the price, but it can safely be said that they are the cheapest in the long run, for one set of discs will out-wear very many sets of knife coulters, and by decreasing the draught (which they do) they save fuel, which alone would cover the extra cost.

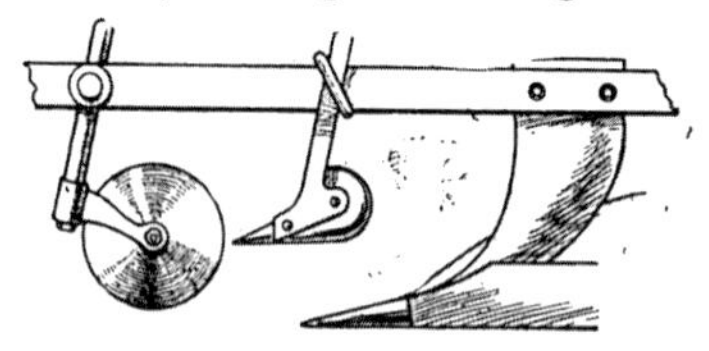

FIG. 3.
Arrangement of skim coulter placed behind a disc coulter.

Disc coulters should be set with the spindle just over the point of the share, and with the bottom of the disc deep enough to make a clean cut, usually about 1½in. above the share. The correct width at which to set them can only be determined by experiment on individual ploughs. A multiple plough never runs dead parallel to its work, but with the beam slightly across the line of travel, with the front end towards the ploughing and the rear towards the land. If the disc be set just in line with the land side of

THE VICTORIAN OIL PLOUGHING ENGINE.

Engine. Two cylinders, 7×8in. (178×203 mm.)

Power. 25-30 h.p. (5,000-6,000 lb., assuming 200 lb. pull per h.p.), 22 b.h.p. at belt pulley.

Fuel. Paraffin.

Weight. 110 cwt.

Gears. Three forward and reverse.

Capacity. Three furrows on heavy land, four furrows on light land; will plough 7-10 acres daily and cultivate 14-20 acres daily when working in pairs with winding drums.

Wheels. Four. Driving wheels, 5ft. × 1ft. 2in.; front wheels, 3ft. 8in. × 8in.

Overall Dimensions. Length 15ft., width 6ft.

the share (which is the correct position for a knife coulter), it will be found in practice that it swings over on its swivel and cuts too narrow. It should therefore be set rather wide, the exact position depending on how much the beam runs out of parallel with the work. The cross set of the plough, referred to above, is caused by the pressure on the tips of the breasts pressing the hind part of the plough towards the land, and no amount of adjustment to the draught rope will overcome it. The only cure would be a very long plough, which is not practicable.

This cross set of the beam is often increased by having the coulters set too wide, or by the hind furrow wheel being worn on its axle, or the hind furrow wheel pillar being very

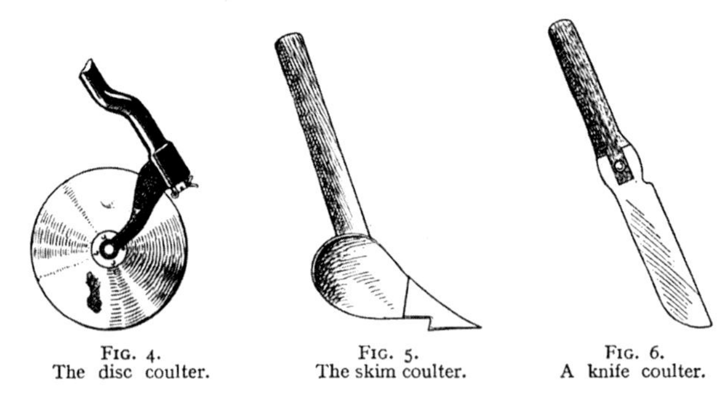

FIG. 4. The disc coulter. FIG. 5. The skim coulter. FIG. 6. A knife coulter.

loose in its bush, any of which faults allow the rear of the plough to swing towards the land side, with the result that the ploughing done is wider than the set width of the plough.

Skim Coulters.

These are a very desirable addition to a plough when breaking up grass land, as they skim off the turf from the upper edge of the furrow slice and drop it in the furrow. The usual way is to set them in front of ordinary knife coulters (fig. 6), but in this position they are apt to pick up the long stuff and "choke" the plough. There is seldom room on a multiple plough to fix them in front of discs, and even when they are so fixed they are no better than when fixed in front of knife coulters. It is not generally known, however, that skim coulters work very well placed behind the discs, the discs being set well forward as in fig. 3. This arrangement permits skims to be used in long grass or weeds, where they would be hopeless placed in front. When arranged thus, the disc should be free to swing from side to side without touching the point of the skim coulter.

Slip Couplings.

Many ploughs have been smashed or hopelessly strained through striking a hidden root or piece of rock, yet many

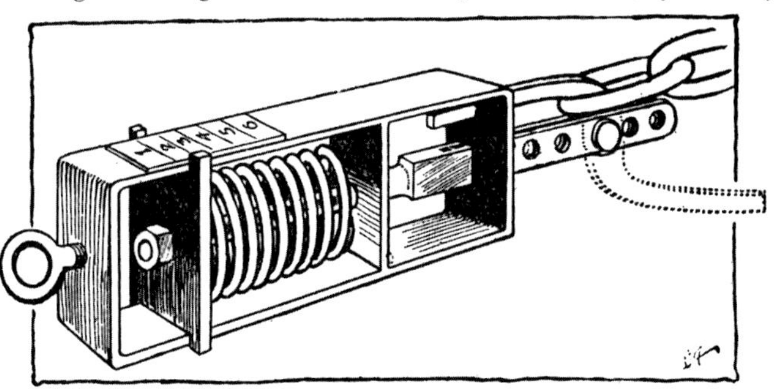

FIG. 7.
A slip coupling which can be adjusted to release the draught rope at a given pressure. When the spring yields to an excessive strain on the draw-bar, the hook is tipped into the position shown by broken lines. The figures on the scale plate on top of the frame denote the predetermined draw-bar pull in tons.

makers (especially British) send out their implements without providing a coupling that will give way under abnormal strain. Several reliable slip couplings are on the market, and two useful ones are shown in figs. 7 and 8. The one illustrated in fig. 7 can be adjusted to release the draught rope at a given pressure, while the chief claim for the other is that, besides being very reliable, it can be

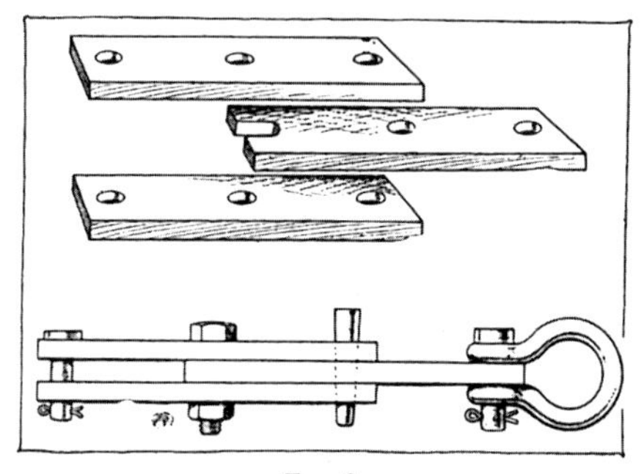

FIG. 8.
A simple form of slip coupling that can be made by any blacksmith.

quickly and cheaply made by any blacksmith; for that reason, its various components are shown in detail.

It consists of three iron plates 7in. × 2in. × $\frac{1}{2}$in., a $2\frac{1}{4}$in. × $\frac{3}{4}$in. steel bolt, and a hard-wood peg slightly tapered and large enough to be a driving fit in a $\frac{3}{4}$in. hole. Each plate is drilled with three $\frac{3}{4}$in. holes, and the end hole in one plate is slotted as shown. All the plates are clamped together by the bolt passing through the centre holes in the outside plates and the slot in the end of the middle plate. The hard-wood peg is driven into the hole which corresponds with the centre of the middle plate, and the device is complete. One end is fixed to the plough and the other to the draught rope by the link and pin shown. If a root or rock is encountered, or any abnormal strain is put upon the draught rope, the wooden peg shears and the slotted end of the middle plate slips away from the clamping bolt.

The Disc Harrow.

Next to the multiple plough, the disc harrow will do more to prove the utility of the motor tractor than any other implement. This fact may take a little time to become apparent over here, for though these implements have been extensively used in Canada and America for many years, they are not well known to the British farmer. Their greatest advantage lies in the fact that they make it possible quickly to reduce coarse wide ploughing to a good seed bed, and it is only because of the difficulty of doing this with the orthodox implements that narrow ploughing is so much in favour in this country. It is a very great advantage when using a tractor to be able to do wide ploughing. Wider wheels can be employed and better adhesion obtained when running in the furrow, and thus the chief difficulty attending the use of the tractor with only one driving wheel (which runs in the furrow) is overcome, viz., its inferior gripping power when doing narrow ploughing. Besides, wide ploughing can be done more quickly and cheaply than narrow ploughing.

The implement illustrated in fig. 9 is a double gang disc harrow with spade-cutting discs front and rear. The usual practice is to have one gang only fitted with spade-cutters and the other gang with plain round ones, and sometimes all plain ones are used. It will be noticed that the discs are saucer shaped, which arrangement causes the soil to be displaced sideways. By setting the discs of the rear gang the opposite way to those of the front gang, the soil is first displaced in one direction and then in the other.

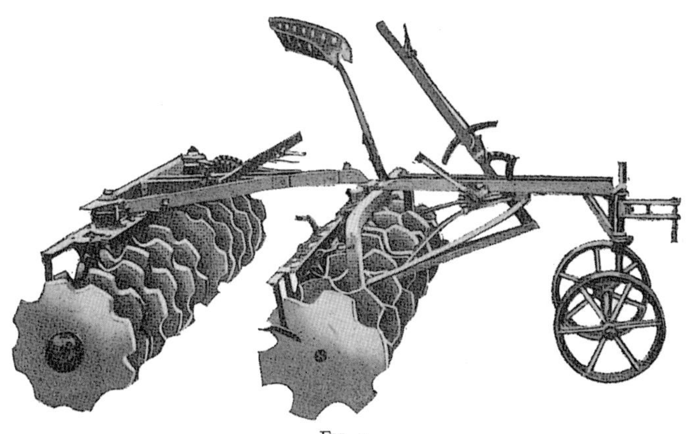

FIG. 9.
A double cut disc harrow with spade-cutting front gangs. The discs on the front gangs are set for out-throw, while those on the rear gangs are set for in-throw.

Just where the old toothed implements fail, the disc harrow is a success. After ploughing tough old ley, a seed bed can be got without danger of turning up the turf, and hard-baked clay can quickly be reduced to a fine tilth when conditions would be quite impossible for toothed implements.

CHAPTER IX.

PLOUGHING.

A TRACTOR travelling at a certain speed, hauling a plough of a given width, would, going in a straight line, plough an acre in so many minutes. The time required will be increased in proportion to the amount of idle running which has to be done when travelling along the headland from one furrow to another, etc., the amount of turning that is necessary, and the time spent in manœuvring when ploughing irregularly shaped pieces. The fuel consumption will be affected in a like degree. It will be obvious, therefore, that when setting out a field provision should be made for reducing idle running and turning to a minimum.

If it were possible to plough right up to the hedge at both ends of a field no headland would be left at all; therefore every yard of headland is so much ground that must be gone over a second time, and represents so much idle running; moreover, as the headland is always compressed considerably by the tractor wheels, the soil is not only harmed, but it may necessitate the use of a two or three-furrow plough to turn it up, whereas a three or four-furrow plough would have done it in the first place. For these reasons a headland should be as narrow as it is conveniently possible to make it. The width necessary will vary according to the type of tractor used, and as a general rule should be about as wide as two-thirds the turning radius of the tractor; that is, if the tractor will turn in a ten and a half yard circle the headland required would be seven yards. A few tractors require a rather wider headland than this, but many a much narrower one.

As a tractor requires a considerable radius in which to turn, it is not wise to finish a field by ploughing the headland backward and forward from end to end, as is the usual practice with a horsed plough, because a lot of time would be wasted when turning at the ends, and a piece of ground equal to the turning radius would be left unploughed in each corner of the field. The correct way is to leave a strip, called a sideland (see fig. 1), down each side of the field, the same width as the headlands, and then finish them altogether by ploughing round and round the field.

Setting Out a Field.

In setting out a field, it may be necessary to consider the position of the old furrows when deciding on the direction and width of the "lands" or sections, but it is not wise to consider the old order too much if a new

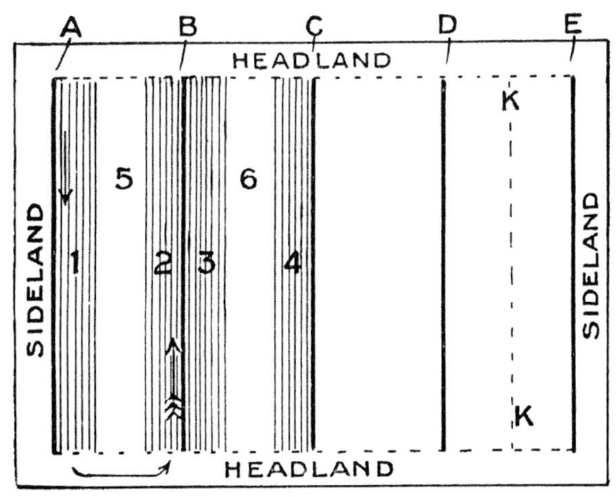

FIG. 1.
Plan of casting and ridging in a rectangular field.

direction or width would aid tractor ploughing, for it is pretty certain the same field will be done more often by tractor than by horses in the future; therefore, any alterations that would be an improvement from the new point of view without interfering with the natural drainage should be made at once.

After the direction of the ploughing has been decided upon, the next thing to do is to mark out the headland the correct distance from the hedge. This aids the ploughman to get the plough in and lift it out at the right spot, and results in a neater finish, which saves time in the end, for it is very often difficult to commence ploughing the headland when the ends of the furrows are very uneven. The marking out of the headland may be done by skimming a shallow furrow with a horse plough, or by the tractor and plough, using only the lever which lowers the hind part of the plough. Another very good way is to have a rigid iron bar fixed to the back of the tractor to which a skim coulter may be clipped, as shown in fig. 2.

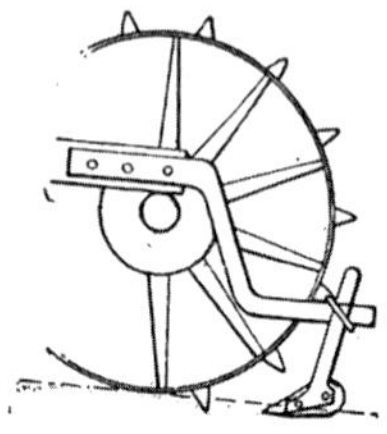

FIG. 2.
Contrivance for marking out a field for ploughing as indicated in fig. 5.

A rectangular field could be ploughed by starting up and down a central line and continuing until the sidelands were reached at either side, but this method would not be economical, as it would involve such a lot of idle running along the headland, especially in a short, wide field, where it would be necessary to do almost as much idle running as actual ploughing.

Starting Furrows.

The correct way is to divide the field into sections of a convenient width by forming starting furrows at intervals. These starting furrows are formed by taking one cut in one direction and one in the reverse direction, throwing the two centre slices together as near as possible. The first two cuts taken in this way form what is termed a "ridge," and the sections between one "ridge" and another are known as "lands." Fig. 1 represents a rectangular field. The ridges are indicated by the heavy lines A B C D E, and the spaces between the "ridges" are the "lands." Sometimes a farmer prefers to set the starting "ridges" with horses, because with a single plough he can split open the ground and then turn it back, in this way leaving no unploughed land beneath the two centre slices which are thrown together. With a multiple plough, a strip of unploughed land a few inches wide is covered over; if, however, the "lands" are fairly wide this does not amount to much in a field, and few farmers mind it. In every other respect a field can be set out as well with a tractor as with horses.

THE WALLIS JUNIOR TRACTOR.

ENGINE.	Four cylinders, $4\frac{1}{4} \times 5\frac{3}{4}$in. (108 × 146 mm.)	CAPACITY.	Three 7in. furrows, and any other farm work where power is required.
POWER.	Constant 10 h.p., maximum 15 h.p. (assuming 200 lb. pull per h.p.)	WHEELS.	Three. Driving wheels, 4ft. diameter × 1ft. face ; front wheel, 2ft. 6in. diameter × 8in. face.
FUEL.	Paraffin.		
WEIGHT.	26 cwt. approx.	OVERALL DIMENSIONS.	Length 11ft. 7in., width 5ft., height 5ft. 4in.
GEARS.	Two forward, one reverse.		

Width of "Lands."

From twenty to thirty yards is a convenient width to have the "lands," the chief consideration being to allow the tractor room to turn when going from one furrow to another without causing unnecessary idle running. In setting out a field, elaborate calculations and exact measurements are to be avoided so long as the "lands" (which should be parallel in a rectangular field) are parallel. Anyone who sets out the "ridges" a distance apart which is a multiple of the width of the plough, in the belief that the ploughing will finish off to a nicety, will be disappointed, for a plough never cuts the exact width it is set to; besides, other influences intervene which would upset the calculations. Upon the straightness or otherwise of the starting "ridges" will depend the neatness of the finished work; therefore some pains should be taken when forming them. The usual way is to make the necessary measurements and to set up guiding sticks at frequent intervals. Another good way, which is infallible after a little practice and saves a lot of time, is to have only one guiding mark, such as a piece of paper, on the hedge at the far end of the field, and a stick or similar mark at the other end from which to start. When the tractor has been started in the right direction, the driver should scan the ground immediately in front of him for some natural guiding mark in the form of a stone or clod or tuft of grass which is immediately between some part of the tractor (say the off front wheel) and the point he wishes to reach. As each mark is reached, another is picked up by the eye, and so on to the end of the journey.

Tractor ploughing cannot be carried out on the same lines as horse ploughing, owing to the greater amount of turning room required by a tractor and plough.

The usual practice with a horsed plough is either to start between two starting ridges—up the left-hand side of E, for instance (see fig. 1), and down the right-hand side of D, and continue (travelling in a left-handed or anti-clockwise direction) until the two ploughed sections meet in the centre of the land at the dotted line K K forming a finished furrow—or to plough up one side of a starting ridge and down the opposite side of the same ridge, travelling in a right-handed or clockwise direction until half the width of the land has been ploughed on either side

THE WALLIS LARGE TRACTOR.

ENGINE.	Four cylinders, 6×7in. (152 × 178 mm.)	CAPACITY.	Four furrows any reasonable depth, and any other farm work where power is required.
POWER.	Constant 20 h.p., maximum 25 h.p. (assuming 200 lb. pull per h.p.), 44 h.p. at belt pulley.	WHEELS.	Three. Driving wheels, 5ft. diameter × 1ft. 8in. face; front wheel, 2ft. 10in. diameter × 1ft. 2in. face.
FUEL.	Paraffin.		
WEIGHT.	76 cwt. approx.	OVERALL DIMENSIONS.	Length 14ft. 3in., width 6ft. 2in., height 7ft. 3in.
GEARS.	Two forward, one reverse.		

of the ridge, so that when the adjoining ridges are ploughed round in the same way the finished sections will meet in the middle of the "lands" and form the finished furrows.

When throwing the work away from a given centre, as in the former method, the operation is termed "casting." When throwing it towards a given centre, as in the latter method, "ridging" is the term used.

As the width of an economical headland does not permit a tractor to make a complete turn, it will be obvious to the reader that neither of the above described methods can be employed, for the tractor could not conveniently turn after the strip of unploughed land, when "casting," had been reduced below a certain width; nor could it turn in the early stages of "ridging," when it would be necessary to travel up and down opposite sides of the same "ridge." The correct way when tractor ploughing is to employ both "casting" and "ridging" in conjunction with each other, which may be done in several ways that will now be described.

Casting and Ridging.

As stated, A B C D E (fig. 1) indicate the starting ridges in a rectangular field, and it is desired to commence ploughing at the left-hand side of the field. "Casting" is employed to commence with, and work is begun along the left-hand side of the ridge B at the point indicated by the feathered arrow, and the return cut taken along the right-hand side of the "ridge" A. This is continued in the direction of the arrows until it is not conveniently possible to turn at the ends, the finished sections being indicated by the groups of parallel lines 1 2. The same operation is now gone through between the "ridges" B C, resulting in the finished sections 3 4. This leaves the two unploughed sections 5 6, which may now be ploughed in conjunction with each other, either by "ridging" round and round 2 and 3 or by "casting" on to sections 4 and 1. If, however, it is desired to have the finished furrows exactly midway between the starting ridges, half of each section 5 and 6 should be done by "ridging," and the other half finished by "casting," or *vice versa*. When following the plan described, the field should be set out into an equal number of "lands," otherwise an odd piece will be left which will be difficult to finish by itself after it has been reduced below a certain width.

Finishing Off.

It is not absolutely necessary to set out all the starting ridges before commencing a field. A very good way is to begin with two only, as indicated at A B (fig. 3). When the sections 1 and 2 have been completed by "casting," the unploughed sections 3 and 4 should be done by "ridging" on to section 2. This will bring the finished ploughing up to the dotted line C. After this, one ridge at a time is formed, and the work continued in the same

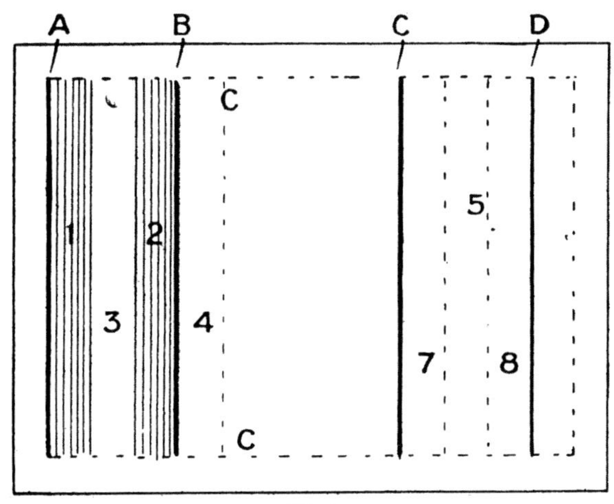

FIG. 3.
Plan of starting ridges in a rectangular field.

manner until the unploughed land remaining in the field only equals the width of about two "lands." This should now be divided down its centre by a starting ridge, so as to form two sections to plough together for finishing off the field. Another way to finish off is to strike the final starting ridge D a distance from the sideland which is equal to the width of section 5, and then, after sections 7 and 8 have been completed, to "ridge" on to section 8 until the section 5 is closed up on the one side and the sideland reached on the other.

It is an interesting fact that, when employing the latter method described of beginning a field with only two ridges set out, the tractor never travels as far along the headland as the width of the "lands" formed between the finishing furrows. For instance, if the two first ridges are thirty yards apart, and each subsequent ridge is formed thirty yards from the finished ploughing, and assuming that 50% of the work is "ridged" and 50% "cast," "lands" forty yards wide will be formed without the tractor ever travelling more than thirty yards along the headland.

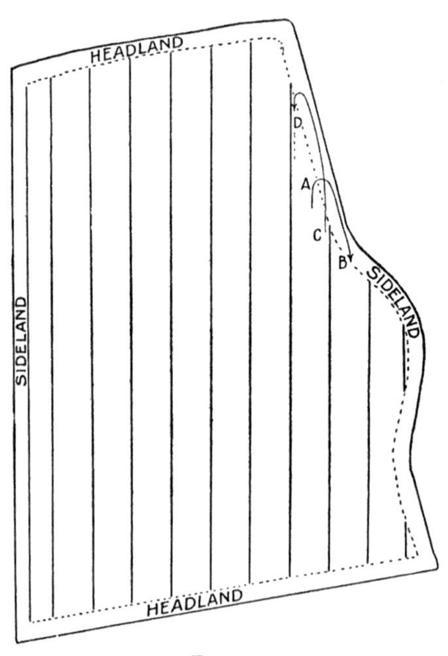

FIG. 4.
Plan showing incorrect method of ploughing an irregularly shaped field.

Irregularly Shaped Fields.

If ordinary intelligence is used and the methods described are carried out, it is not possible to waste much time in the form of unnecessary light running in a rectangularly shaped field, but, when setting out an irregular field, to reduce idle running and turning to a minimum a little forethought is necessary.

The direction of the ploughing in such a field has a great influence on deciding not only the amount of turning and idle running that must be indulged in, but the width of the headlands and sidelands as well. Fig. 4 shows an irregularly shaped field, the parallel lines indicating the plan of setting out usually adopted for horse ploughing, and unfortunately, and to the detriment of the land and at a great sacrifice of time and fuel, the same plan is often followed by inexperienced tractor ploughmen.

It will be noticed that many of the "lands" run out to the sidelands at an acute angle; therefore, when following such a plan, the tractor must make almost a complete turn on the sidelands at the end of each round when ploughing these "lands." Almost a complete turn would have to be made, for instance, when leaving A to set in again at B, or when setting in at D after leaving C. This necessitates leaving a wider sideland than usual, and a headland of equal width to plough in conjunction with it; besides, the distance between the ends of the furrows is increased by the acuteness of the angle at which they run out to the sidelands. This means that a lot of idle running is inevitable, and the wide sideland is badly "packed" by the number of times the tractor must pass over it.

The Correct Way.

The most economical way to plough such a field is to follow the plan illustrated in fig. 5. Here turning on the sidelands is avoided, excepting where the hedge bulges out into a semicircle, because the adjoining "lands" on either side of the field are parallel to them. By this method one or more of the "lands" must be "piked" or tapered, and in the diagram the two centre ones are so shaped. Before the field can be finished in the ordinary way these tapered "lands" must be got parallel, and this is accomplished in the following manner. "Casting" is employed, and when the narrow end is of a width which just permits the tractor conveniently to turn, the succeeding furrows are made shorter and shorter each round by lifting out the plough and setting it in on the opposite side of the "land," as indicated by the parallel lines 1 2 and curved arrows T T. When the tapered "lands" have been got parallel in the manner described, one or two complete turns will suffice to straighten up the ragged edges where the plough has been lifted out and set in. After this the procedure will be the same as for ploughing a square field.

To show the ploughman exactly where to lift out and set in the plough when making the short turns, the contrivance shown in fig. 2 is useful to scratch two parallel marks between the points M M.

Ploughing up to a Curved Hedge.

On the right-hand side of the diagram (fig. 5), it will be noticed that the hedge bulges out into a semicircle. Such a place is often met with, and gives an inexperienced ploughman no end of trouble. The correct way to do it is to tackle the semicircular piece separately and before the adjoining ploughing is started. The sidelands should be marked out, and a ridge S S should be set midway down what remains of the semicircle or bulge. Shorter and shorter furrows (3) should be ploughed on the side nearest the hedge, and longer and longer ones (4) on the opposite side of the ridge until the bulge is closed up. A certain amount of idle running is inevitable, and complete turns must be made at each end; hence the necessity of doing such a piece before the adjoining ploughing is commenced.

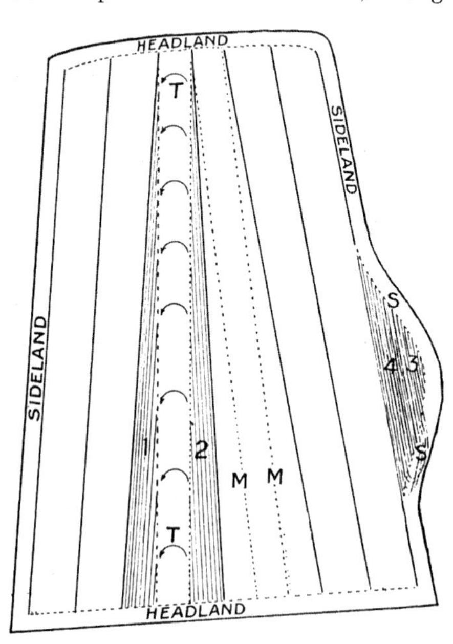

Fig. 5.
Plan showing most economical method of ploughing an irregularly shaped field.

Curved "Lands."

The "lands" in many of the old grass fields which have not been ploughed for years are often both winding and tapered in a manner that is not at all adapted to tractor

ploughing. In many instances the ridges are so high and the furrows so deep that the direction cannot easily be altered, because a tractor plough will not cut across pronounced ridge and furrow without digging in too deeply on the former and slipping out in the latter. In a few cases it will not be considered advisable to alter the direction of the old "lands" on account of the drainage, but where it is desirable to alter the direction from winding to straight and from tapered to parallel, and the ridge and furrow are too pronounced to permit this being done, the best plan to follow is to do the ploughing in the old direction for the first or even the second year, each time forming the starting ridges in the old furrows and the finishing furrows on top of the old ridges. This in time will get the surface flat enough for the "lands" to be altered in any direction desired.

Ploughing the Headlands and Sidelands.

The headlands and sidelands should be begun so that the last round will finish near the gate.

As a rule, it is best to travel in a right-handed direction, throwing the work towards the field and finishing by the hedge; otherwise the last rounds will be made with the wheels on the ploughing. If, however, the field is subsequently to be cross-ploughed, it is best to start by the hedge and then turn the work back, by doing it in the reverse direction after cross-ploughing. The plough should only be left in when taking the corners if the soil is light. If it is heavy or hard, a strain is put on the plough which often results in broken breasts and coulters.

If it is found difficult to get the first cut along the headlands on account of the ragged edges of the furrows, a good plan is to run backwards and forwards once or twice and roll them down with the tractor wheels without the plough.

It is when nearing the hedge that precautions must be taken against striking hidden roots.

Cross Ploughing.

This is an operation that can be performed really well by a tractor and multiple plough. The condition of the ground should be fairly dry, or owing to the weight of the tractor the soil will come up so stiff as to be in no better condition than after the first ploughing. The most suitable tractor to use is one with two wide driving wheels, or, if with narrow driving wheels, they should be fitted with extension rims shod with extra grips.

The wider wheels or addition of extension rims give a better grip, lower the degree of soil compression, and check that tendency of partial slipping or floundering which is caused by the looseness of the surface. They also have a rolling effect upon loose soil, which, without unduly compressing it, aids the plough to cut through without getting "choked up." Blocking or "choking" of the plough through loose, tough, or weed-infested soil driving up in front is the chief difficulty when cross-ploughing with a multiple plough; therefore the greatest advantage should be taken of the rolling effect of the tractor wheels by running with all the wheels out of the furrow. Another reason why the tractor wheels should not run in the furrow when cross-ploughing is that, the soil being loose, it is impossible to cut a clean furrow, for a lot of soil falls back, partially filling it. If this is run over and pressed down by the tractor wheels, though it will not show when covered over by the succeeding furrow slice, the damage caused will be noticeable in the ultimate crop.

The American type plough with wide bottoms and short breasts is excellent for cross-ploughing, and will often do good work where a narrow plough with long breasts would fail. If an adjustable plough be used, it should be set to its greatest width. A plough that has plenty of space between the ground and the beam is also an advantage, and this should be borne in mind when purchasing an implement.

Coulters.

Knife or stem coulters are of very little use for cross-ploughing; either discs should be used, or none at all. The unfortunate thing about coulters of any kind on a multiple plough is that each one comes in line with the preceding frame or body, partially closing up the passage, and it is often found, when the plough has a tendency to choke, that by removing the coulters altogether the soil will break against the "neck" and pass through quite easily.

Other Cultivation Operations.

The tractor has not as yet been used extensively in this country for anything but ploughing, and the small mixed farmer would like to know with what success it takes the place of horses in other respects, and to what extent he can use his old horse-drawn implements (which he cannot afford to scrap), such as cultivators, harrows, rollers drills, reapers, and binders.

THE WEEKS-DUNGEY "NEW SIMPLEX" TRACTOR.

ENGINE. Four cylinders, $3\frac{3}{4} \times 5\frac{1}{4}$in. ($95 \times 133$ mm.)

POWER. $22\frac{1}{2}$ h.p. at 1,000 r.p.m. and governed.

FUEL. Paraffin.

WEIGHT. 35 cwt.

GEARS. Three forward, one reverse.

CAPACITY. Two to three furrows according to soil, two binders or mowers, 3-4 tons on the road.

WHEELS. Four, two tracks. Driving wheels, 3ft. 4in. diameter × 10in. face; front wheels, 2ft. 6in. diameter × 6in. face.

OVERALL DIMENSIONS. Length 8ft. 2in., width 4ft., height 5ft. 6in.

It is the farmer who has used tractors the longest, and who has been the most enterprising in the matter of experiments, who is the most enthusiastic regarding the utility of a tractor for these jobs. No doubt larger and heavier implements, especially made for use with a tractor, do better work and are cheaper in the long run than the lighter horse implements, for it must be borne in mind that a medium-powered tractor has the hauling capacity of nine or ten horses, whilst the old implements are designed to be hauled by from two to five horses. However, a lot can be done by coupling up two or more implements together in such combinations as a drill followed by a pair of harrows, a cultivator and a crusher, or two or more binders. Again, a tractor capable of hauling two four-horse implements at the same pace as a team of horses (about 2½ m.p.h.) would, on a higher gear, haul a single implement at 5 m.p.h., and do the same quantity of work at about the same cost per hour for fuel and labour as when hauling two.

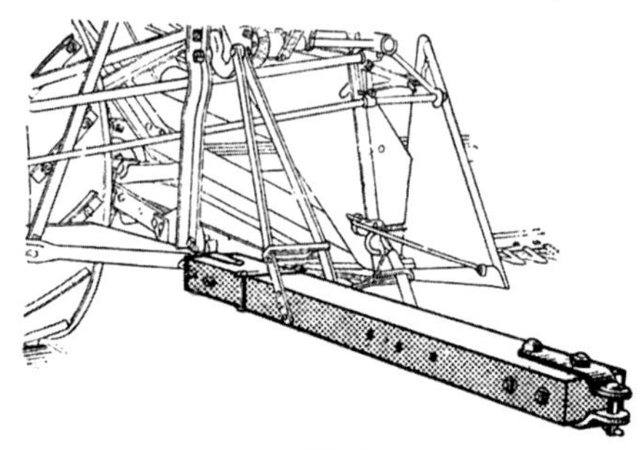

FIG. 6.
The McCormack stub-tongue hitch for one binder alone, showing the clevis on the stub-tongue.

Hitches and Couplings.

Regarding coupling up various horse-drawn implements to a tractor, the owner would be well advised to consult his implement maker, and get his advice on the matter of chains, draw-bars, stretcher-bars, etc.

For drawing a single binder or mower, nothing more is necessary than a short stub-pole fitted with a draught eye or other means of flexibly connecting it to the tractor draw-bar (see fig. 6). Many ingenious hitches have been designed, however, which give better results at the corners. The Whiting-Bull hitch (fig. 7) is a good example.

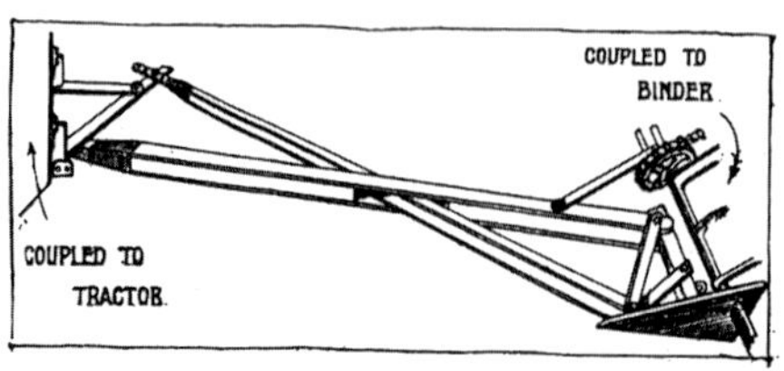

FIG. 7.
The hitch device marketed by the Whiting-Bull Co.

WAKEFIELD
AGRICASTROL
MOTOR OIL
The
ONE OIL
for Motor Tractors
Just one illustration of
"Agricastrol" quality —
The Board of
Agriculture
for Scotland
have placed a
contract for
"AGRICASTROL."
VERB. SAP.
Full particulars on request.
C.C. WAKEFIELD & Co. Ld
[Sir Chas. Wakefield, Bart.,
Governing Director].
Wakefield House
Cheapside, London
E.C.2.
Immediate delivery to
rails guaranteed
AGRICASTROL
MOTOR OIL
ROL
OIL
C.D.C.

For coupling up two or more binders, a special hitch is required to ensure cutting an even swath with each machine and to enable the corners to be taken squarely. Fig. 8 shows a contrivance marketed by the Overtime Co. which, they claim, answers the dual purpose. It will be seen that the binders are coupled up stepwise, so to speak, and the steps are adjusted and prevented from opening out by the steel cable B B. The method adopted by the International Harvester Co. is to hitch the first binder to the tractor by an ordinary strengthened stub-pole, and to couple the second and each subsequent binder to the preceding machine by a stub-pole the draught angle of which is adjustable at will by means of a quadrant and pinion arrangement under the control of the man on the seat. Fig. 9 shows the rear of a binder and the method of attaching the steering stub-pole of the following machine; whilst fig. 10 shows the steering arrangement referred to.

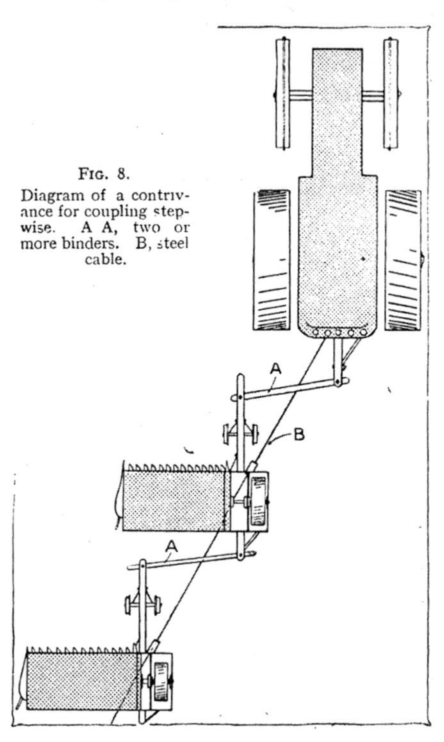

FIG. 8.
Diagram of a contrivance for coupling stepwise. A A, two or more binders. B, steel cable.

Fig. 11 is a plan view of the International Harvester Co.'s hitch for two grain drills. A similar arrangement to this suggests itself as a very suitable one for coupling up other implements such as a pair of rollers or cultivators.

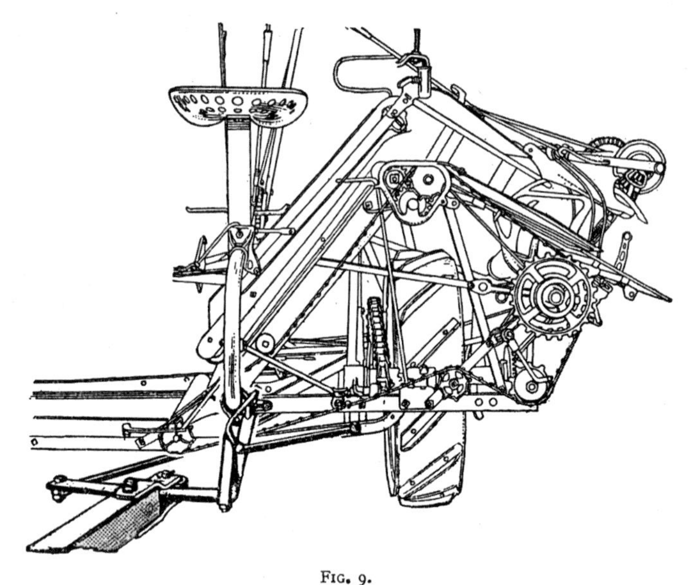

FIG. 9.
The International Harvester Co.'s method of coupling two binders to a tractor. Note the steering-tongue hitch at the rear of the first binder.

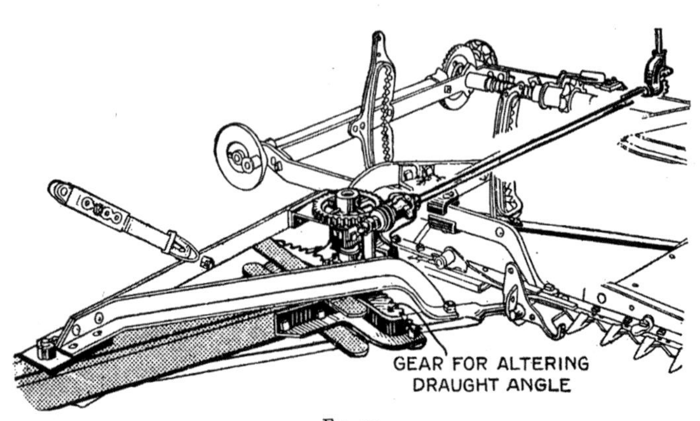

FIG. 10.
The Deering steering-tongue hitch, which enables the driver to alter the draught angle without leaving his seat.

Belt Work.

The engine as employed in any good make of tractor is particularly suitable for doing any kind of belt work within its power, such as threshing, baling, pumping and sawing, and for turning the various small machines to be found on any farm, such as pulpers, grinders, chaff-cutters, etc. The chief advantage of using the tractor for these jobs is that a single capital outlay purchases the whole power plant for the farm. If the work is well arranged, the engine will be usefully employed a great proportion of the year, and the farmer will quickly see his money back in the form of profitable service. The ease with which a tractor engine can be started and the despatch with which it can be moved from one job to another are other advantages.

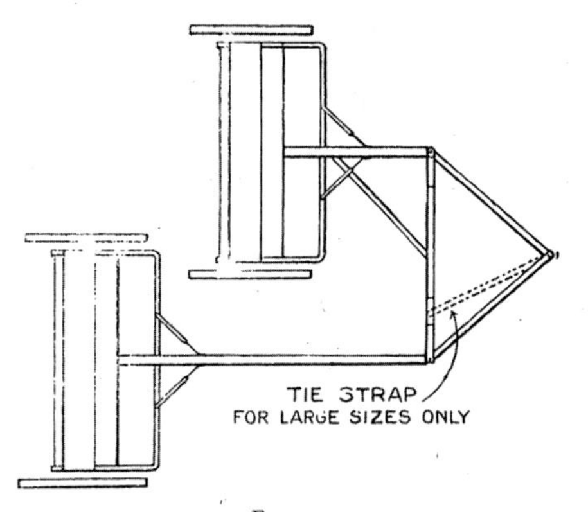

FIG. 11.
Plan view of triangular hitch for grain drills.

Threshing is the heaviest belt work that would be demanded of a tractor engine; 25 h.p., which is a very desirable power for ploughing, will drive a full-sized threshing drum with ease. The same power would be ample to turn a whole lot of smaller machines, such as pulpers and grinders, etc., at one and the same time, and it would not be economy to drive them singly. A line of shafting, fitted with a fast and loose pulley for each machine, should be put up. All the machines could thus be working at once, or any number of them be cut out by passing the belts on to the loose pulleys.

If the prospective purchaser of a tractor has in view the doing of much belt work, before settling on a machine he should assure himself that it is fitted with an efficient governor, and ascertain the position of the belt pulley.

The former is essential if constant speed and steady running are required; and if the latter is placed in the wrong place it will cause no end of inconvenience.

Pulley Speed.

The size and speed at which the belt pulley runs are also important considerations. With a small pulley a slipping belt is inevitable, yet absurdly small ones are often found on otherwise good tractors, particularly on those employing vertical high-speed engines with the pulley taking its drive direct from the engine-shaft, the reason being that

This is the latest model Fordson tractor and differs only in detail from the one illustrated on page 111. The specification is as follows :

ENGINE.	Four cylinders, 4 × 5in. (102 × 127 mm.).	WEIGHT.	22 cwt.
POWER.	A draw-bar pull of 1,800 lb. at ploughing speed and in low gear of 2,500 lb. 22 h.p. at belt pulley at 1,000 r.p.m.	CAPACITY.	Three furrow plough, genera farm work and stationary work. Road speed five miles an hour.
FUEL.	Paraffin, starting on petrol.	WHEELS.	Four. Driving wheels 3ft. 6in. diameter × 12in. face.
GEARS.	Three forward, one reverse.	WHEELBASE.	5ft. 3in., tread 3ft. 2in.

a comparatively large pulley revolving at 700 to 1,000 r.p.m. gives too quick a drive for ordinary purposes. The obvious cure is to employ a large pulley driven through a reducing gear, and this method is adopted on many machines.

Engines running at 700 to 1,000 r.p.m. are usually of the four-cylinder vertical type set in line with the main

frame, and to get the pulley on the right or left side of the machine—which is the correct position—a reducing gear embodying a right-angle drive is usually employed. On an engine set across the frame and running at 300 to 500 r.p.m., as do the majority of single and twin-cylinder horizontal engines, there is, of course, no better position for the belt pulley than on the end of the crankshaft, or driven direct therefrom through the clutch.

Pulley Position.

The belt pulley may be found in three different positions: on the side of the machine, parallel to the frame; on the end of the crankshaft in front of the machine; or on a tailshaft at the rear—in the two latter positions running at right angles to the frame. The first is the best, as when in this position the machine can be driven up to its work and moved backwards or forwards to get the right tension on the belt. If, however, the pulley is placed fore or aft, the tractor must necessarily stand at right angles to its work. This position not only makes it difficult to set the machine so as to get the right tension on the belt, but it will be found very difficult to set the machine at all where space is restricted, such as in a building or a rickyard.

A paraffin engine cannot be readily stopped and started like a steam engine; therefore provision must be made for quickly disconnecting the power from whatever is being driven without stopping the engine. When the clutch is placed between the engine and the pulley, this is done by simply putting out the clutch. In other cases a pulley is employed embodying a separate clutch, but in many instances the belt pulley is rigidly fixed to the crankshaft. Some machines are not fitted with a belt pulley at all, while others are minus a governor: so it behoves the prospective purchaser of a tractor to have a good look round before making a choice.

Example of a Belt Drive Layout.

The following example of a belt drive layout for a farm where a tractor is used has been supplied to us by Messrs. Henry Garner, Ltd., the makers of the Garner tractor. As will be seen on reference to the accompanying drawing, the implements employed are a grinding and crushing mill, a pulper and root cleaner, a chaff cutter, and a cake breaker. The horse-power of the Garner is 29 h.p., the size of tractor pulley 8in. diameter by 8¼in. wide, and the governed speed

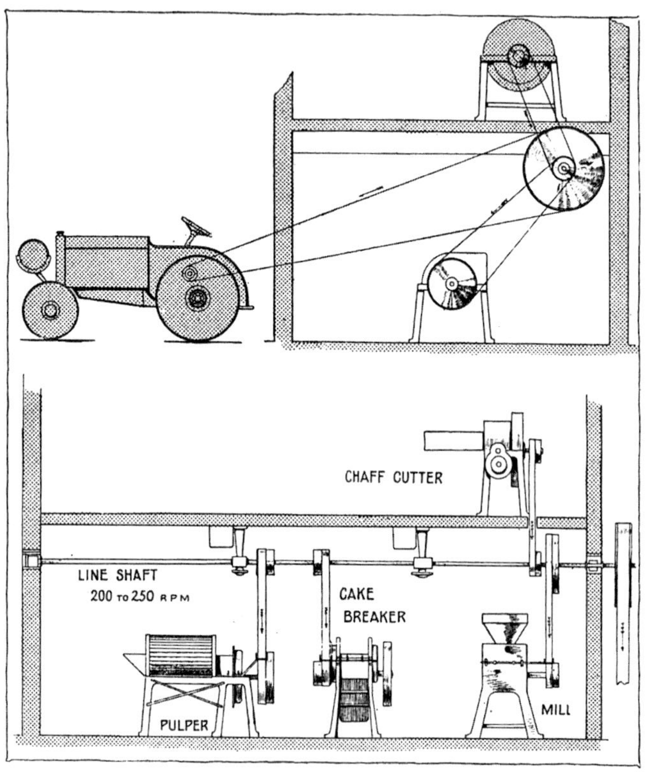

LAYOUT OF LINE SHAFT, PULLEYS, AND AGRICULTURAL IMPLEMENTS WHICH CAN BE DRIVEN FROM THE TRACTOR PULLEY.

The arrangement shown is the one recommended by the makers of the Garner tractor. See text for detailed explanation.

of the engine 1,000 r.p.m. Other examples are usually obtainable from other tractor makers or agents on application.

In setting out the machines and providing shafting, provision should be made for future extensions.

Where the machines can be grouped, and the engine has ample power, it is advisable to drive through a line shaft, in order to economise time and fuel.

The usual speed for a line shaft is 200 to 250 r.p.m., and for this power and the length required a shaft of $1\frac{1}{2}$in. to $1\frac{3}{4}$in. diameter should be used.

Take 200 revs. per minute and size of shafting $1\frac{3}{4}$in.

Therefore, size of main drive pulley on line shaft ... ... $= \frac{8 \times 1000}{200} = 40$in.

Width of pulley on account of using same belt as for thresher should be = 7in., and should have a curved face.

Machines to be Driven.

(1.) *Chaff Cutter:*

Speed	200 revs. per min.
Fast and loose pulleys ...	12in. dia. × $4\frac{1}{2}$in. face
Power required	3 h.p.

(2.) *Pulper and Root Cleaner:*

Speed	90 revs. per min.
Fast and loose pulleys ...	26in. dia. × $3\frac{3}{4}$in. face
Power required	2 h.p.

(3.) *Cake Breaker:*

Speed	120 revs. per min.
Fast and loose pulleys ...	18in. dia. × $3\frac{1}{2}$in. face

(4.) *Grinding and Crushing Mill:*

Speed	500 revs. per min.
Fast and loose pulleys ...	14in. dia. × $5\frac{1}{2}$in. face
Power required	8 h.p.

Sizes of Pulleys on Line Shaft to Drive—

Chaff Cutter $= \frac{200 \times 12}{200}$ = 12in. dia. × 9in. flat face.

Pulper $= \frac{90 \times 26}{200}$ = 12in. dia. × $7\frac{1}{2}$in. flat face.

Cake Breaker $= \frac{180 \times 120}{200}$ = 11in. dia. × 7in. flat face.

Mill $= \frac{500 \times 14}{200}$ = 35in. dia. × 11in. flat face.

CHAPTER X.

STEAM TRACTORS.

STEAM is the power by which during the past century a social and industrial revolution has been created surpassing in magnitude the material progress recorded during the previous thousand years, and though successful internal combustion engines were running in the seventies, they did little more than supplement steam in a humble way until the end of the century. Since then, the rapid progress made by the internal combustion engine is probably the greatest romance of the engineering industry, for in less than twenty years it has revolutionised road transit, conquered the air, and at the present day is supplanting steam in many directions. Steam during the past sixty years has done much for the farmer, and it is not unnatural that those who have had such good service from it should ask: What are the virtues of the internal combustion engine, how does it compare with steam, and what part is steam likely to play in the modernising of agriculture which is just beginning to take place?

Internal and External Combustion.

Both the steam engine and internal combustion engine are heat engines; that is, the latent heat in the fuel is released by combustion and converted into motion. In the case of the steam engine, the burning fuel in the furnace expands water into steam in a closed chamber (the boiler), and the resultant pressure set up is utilised to drive a piston backwards and forwards in a cylinder, the reciprocating piston acting upon suitable connecting rods which turn a crankshaft. In the case of the internal combustion engine the carburetter prepares a mixture of fuel vapour and air, which is admitted directly into the cylinder and there ignited, the expansion due to the resultant explosion or very rapid burning of the charge having the same effect upon the piston as steam in a steam cylinder. It will be seen, therefore, that the functions performed by the very

THE

SAUNDERSON UNIVERSAL TRACTOR

For Ploughing, Threshing, Reaping, Cultivat- :: ing, Driving Machinery, and Transport. ::

23-25 h.p.—Model G.

As supplied to H.M. The King, H.R.H. The Prince of Wales, The War Office, Food Production Dept., etc.

SAUNDERSON TRACTOR PLOUGHS

2, 3, and 4 Furrows.

Strong, reliable, independent type general purpose ploughs equipped with every movement necessary for Tractor work.

PROMPT DELIVERY FROM STOCK.

Full Supply of Spares for Tractors and Ploughs kept.

Write us for full particulars.

THE

SAUNDERSON TRACTOR & IMPLEMENT CO., LTD.,

Elstow Works, BEDFORD, Eng.

MANN'S STEAM TRACTOR.

ENGINE. Two cylinders, 4 and $6\frac{3}{8} \times 8$in. (102 and 162×203 mm.)
POWER. Will pull 6 tons up a steep hill, 25 h.p. at belt pulley.
FUEL. Coal.
WEIGHT. 90 cwt.
GEARS. Three forward and three rever .

CAPACITY. Three furrows in strong land, drive a 4ft. 6in. thresher with a straw elevator, and haul 6 tons on the road.
WHEELS. Four. Back, 4ft. 3in. diameter × 1ft. 8in. face; front, 2ft. 11in. × 8in. face.
OVERALL DIMENSIONS. Length 13ft. 2in., width 6ft.

heavy, bulky, and expensive boiler and furnace, and the manual labour involved in stoking in the one case are in the other entirely performed by the comparatively tiny, light, and inexpensive carburetter. Moreover, the heat

MANN'S STEAM TRACTOR (FRONT VIEW).

J. Reversing lever. K. Brake handle. O. Steerin gwheel.
P. Steering chain quadrant. R. Front axle springs

generated in the internal combustion engine—being right on top of the piston—can be employed to the very best advantage; whereas in the steam engine much of it is lost

Designed for Maximum Power.

Reserved at present for work of urgent National importance.

Illustrated Catalogue post free from the makers:

THE LODGE SPARKING PLUG CO., LTD., RUGBY.

owing to the indirect manner of its application, the greatest loss taking place up the chimney in the form of gas and smoke, which, of course, represents valuable unburnt fuel. This explains what is meant when it is said that the internal combustion engine has a higher fuel efficiency than the steam engine. The former makes better use of the amount of heat available in a given quantity of fuel, but it does not

Rear View of **MANN'S STEAM TRACTOR** showing Transmission.
A. Engine shaft. B, C, D, E. Gear wheels. G. Winding drum.
I. Trailer draw-bar. M. Plough draw-bar.

necessarily indicate that the latter is the more expensive to run; that depends chiefly on the relative price of their respective fuels.

The rapid development of the internal combustion engine is not, then, the result of lower fuel cost, but because, by substituting the carburetter for the furnace and boiler and using liquid fuel in place of solid, we get a compact and absolutely self-contained power plant that is incomparably lighter and smaller than a steam plant of equal power. Therefore, when lightness in relation to power is the chief consideration—as in aircraft, for instance—the internal combustion engine has no rivals. On the other hand, where weight is immaterial, or advantageous as with the railway engine, steam still holds its own. It would appear, therefore, that for stationary work, for double and single engine cable work, and for tractor work where weight is a necessity

THE SUFFOLK PUNCH STEAM TRACTOR.

or not undesirable, there is still a sphere of usefulness for the steam engine on the land, which will be limited or enlarged in the future chiefly by cost of construction, the relative cost of fuels, and the general trend of tractor design. If the trend of design is in the direction of the light powerful machine, it seems unlikely that steam will be able to compete successfully. The greater elasticity of steam as a power medium, however, and the form of valve motion

generally employed, give the engine a much wider range, both in power and speed, and owing to this the transmission can generally be made simpler.

THE SUMMERSCALES STEAM TRACTOR.

ENGINE.	Four cylinders, V type, 4 × 7in. (102 × 178 mm.).	GEARS.	Camshaft reducing gear, single-speed.
POWER.	25 h.p. at 350 revs. per min.	CAPACITY.	Four-furrow plough. Stationary work. Road speed five miles an hour.
FUEL.	Coal or coke, 60 lb. to the acre on average land.	WHEELS.	Three. Back, 4ft. 6in. diameter × 1ft. 9in. face.
WEIGHT.	65 cwt.		

OVERALL DIMENSIONS Length 12ft. 4in. width 5ft. 10in., height 8ft. 1in.

CHAPTER XI.

CONVERTING A TOURING CAR INTO A TRACTOR.

MORE than one enterprising firm have brought out a tractor attachment which, when fitted to an ordinary Ford car, converts it into an efficient tractor that has a capacity for work equal to many self-contained tractors costing much more money. Anything in the form of a make-shift is naturally looked upon with suspicion, but the make-shift in this case is so very practical that it is well worth considering, especially by a farmer who owns a Ford which is giving little or no service owing to the scarcity of petrol.

To mention one make that has "made good," the Eros attachment consists essentially of two tractor wheels and an axle, with the necessary fittings for connecting them to the Ford car. The back wheels of the car are taken off and the attachment is bolted to the frame a little behind the ordinary axle. In place of the back wheels which have been removed, two small driving pinions are fitted to the ends of the axle proper of the car, and these pinions mesh with master gears in the form of internal toothed rings on the tractor wheels. The pinions employed on the ends of the axle proper usually have six teeth, whilst the master gears on the tractor wheels each have sixty-six teeth; thus the axle speed of the tractor is only one-eleventh the axle speed of the touring car at a given engine speed. It does not follow, however, that the speed of the tractor combination is only one-eleventh that of the car, for the tractor wheels are 38in. in diameter, the car wheels being but 30in. The actual speed of the tractor works out at about 2½ m.p.h. when the engine is revolving at a speed that would drive the car at 20 m.p.h. on high gear, which, by the way, is the right gear to use, the lower gear being reserved for emergency. As the turning power of the wheels increases almost proportionately to the gear reduction obtained by the pinions on the axle proper, the master gears, and the size of the tractor wheels, it follows that the draw-bar pull of the vehicle is increased nearly eight times by the use of the attachment, the power obtained being ample for hauling

THE EROS ATTACHMENT.

This is a Device for converting a Ford Car into a Tractor, and is shown pulling a Plough.

a two-furrow plough through fairly stiff land, and more than sufficient for performing the lighter cultivating operations and for drawing a binder or reaper.

Hauling a Trailer.

By employing a special trailer, the combination can also be made to haul a heavy load along the highway. The trailer has two wheels only placed towards the rear, its fore end resting on a suitably arranged cross-piece or bolster above the tractor axle, so that some of the weight of the load is transferred from the trailer to the tractor wheels, which has the effect of giving them greater gripping power. This is a principle that is sure to be heard more of in the near future, for it is the necessity for great weight in a road tractor that makes it in many cases unsuitable for the land.

One usually associates a tractor with titanic strength and solid construction, and it is not to be wondered at that the untechnical mind should consider an ordinary Ford car too frail to stand the stress of tractor work. As a matter of fact, very little extra strain is put upon any part of the car by the conversion. The strain on the mechanism can never exceed that which is occasioned when the engine is developing its full power, and if it were unable to stand this it would be useless as a touring car. The only trouble that has been met with is that of overheating, due to the slower rate at which the radiator meets the air and the demand for a more constantly sustained throttle opening and engine speed when ploughing.

Cooling.

The difficulty has been overcome, in the case of the particular attachment described, by fitting a larger radiator and fan and the provision of a simple water circulator to aid the thermo-syphon system of cooling. The circulator in question consists of a small bore pipe led from the exhaust manifold, ending in a nozzle inside the bottom water pipe and pointing in the direction of the water flow. As the exhaust gas issues from the nozzle, it carries the water with it, so to speak, and accelerates the circulation. A forced feed lubricating system is included, which pumps oil through a sight feed to all the engine bearings, and replaces the usual and less positive splash feed.

LUBRICATES MOST
THE PERFECT LUBRICAN
W. B. DICK
90, FENCHURC
Telephone: AVENUE 7854 (4 lines).

ARBONIZES *LEAST*

OR FARM TRACTORS.

& Co., Ltd.,

STREET, E.C.3.

Telegraphic Address: "DICOTTO, LONDON."

THE WHITING-BULL TRACTOR. An American Machine with Single Front Steering Wheel in Line with the Driving Wheel. The Land Wheel runs loosely on its Axle. Specification and Description on Pages 116-118.

ADDENDA.

Amended detailed specifications of the following tractors were compiled after Chapter VII. had gone to press :

ALLDAYS.

ENGINE.—Four cylinders, 4 × 5$\frac{1}{8}$in. (102 × 130 mm.), 30 h.p. at 1,000 r.p.m.
POWER.—30 h.p., draw-bar pull 2,000 lb.
FUEL.—Paraffin, starting on petrol.
GEARS.—Three forward (1$\frac{1}{4}$, 2$\frac{1}{4}$, and 5 miles per hour), one reverse. Final drive by single roller chain.
WEIGHT.—55 cwt.
CAPACITY.—Three furrow plough, road haulage, threshing, etc.
WHEELS.—Four. Drivers, 5ft. × 12in. face.
BELT PULLEY.—On off side, driven by bevel gearing, and under control of main clutch.
GOVERNOR.—Centrifugal.
BALLAST TANK.—Over rear axle, capacity 1,200 lb. of water.
SPRINGS.—Central coil spring to front axle and two-leaf springs to rear axle. Latter can be locked out of action.

BURFORD-CLEVELAND.

ENGINE.—Four cylinders, 3$\frac{11}{32}$ × 5$\frac{1}{8}$in. (85 × 130 mm.), 20 h.p. at 1 000 r.p.m.
FUEL.—Paraffin, petrol for starting.
GEARS.—One forward (about 3 to 5 miles per hour), one reverse. Final bevel drive to rear axle.
WEIGHT.—27 cwt.
CHAIN TRACKS.—Double track, each 4ft. 2in. × 6in. wide.
STEERING.—Mechanical through brakes on differential shafts.
BELT PULLEY.—On extension of crankshaft, not under control of clutch.
GOVERNOR.—None.
COOLING.—Centrifugal pump, radiator, and fan.

OVERTIME.

ENGINE.—Two cylinders, 6 × 7in. (152 × 178 mm.), 24 h.p. at 500 r.p.m.
POWER.—12 h.p. (2,400 lb., assuming 200 lb pull per h.p.), 24 h.p. at pulley.
FUEL.—Paraffin.
WEIGHT.—43 cwt.
GEARS.—One forward (2$\frac{1}{2}$ miles per hour), and one reverse. Final drive by exposed pinions to rear wheels.
CAPACITY.—Four furrows, full size thresher, etc.
WHEELS.—Four. Driving wheels, 4ft. 4in. × 10in. face.
BELT PULLEY.—On near side, controlled by main clutch.
GOVERNOR.—Exposed centrifugal type.
COOLING.—Centrifugal pump, radiator, and fan.
OVERALL DIMENSIONS.—Length 11ft. 10in. × 6ft. 1in.

WHITING-BULL.

ENGINE.—Two cylinders, 5$\frac{1}{2}$ × 7in. (140 × 178 mm.), 24 h.p. at 750 r.p.m.
POWER.—12 h.p. (2,400 lb., assuming 200 lb. pull per h.p.), 24 h.p. at pulley.
FUEL.—Paraffin. WEIGHT.—44 cwt.
GEARS.—One forward (2$\frac{1}{2}$ miles per hour), and one reverse. Final drive by roller pinion to gear on driving wheel.
CAPACITY.—Three furrows, light land four furrows ; general agricultural work.
WHEELS.—Three. Front wheels, 2ft. 6in. × 6in. ; centre rib, 3in. ; land wheel, 3ft. 4in. × 8in. ; main drive wheel, 5ft. × 12in. or 14in.
BELT PULLEY.—12in. diameter × 6$\frac{1}{2}$in. wide, on left-hand side, and controlled by separate clutch.
GOVERNOR.—Enclosed, running in oil.
COOLING.—Pump, radiator, and fan.
OVERALL DIMENSIONS.—Length 13ft. 7in., width 6ft. 6in., height 6ft. 6in.

WYLES.

ENGINE.—Single-cylinder, 5 × 6in. (127 × 152 mm.), 11 h.p. at 1,000 r.p.m.
FUEL.—Petrol or Wyles's fuel oil.
GEARS.—Two forward (1$\frac{3}{4}$ and 2$\frac{1}{4}$ miles per hour), one reverse. Final drive by enclosed gearing to driving wheels.
CAPACITY.—Two furrow plough ; belt pulley provided for light stationary work ; an attachment can be supplied to convert the machine to a tractor.
WEIGHT.—21 cwt., with two furrow plough.
WHEELS.—Two driving wheels only, 2ft 9in. × 7in. face.
BELT PULLEY.—Off side, clutch controlled.
GOVERNOR.—Centrifugal, enclosed and running in oil.
COOLING.—Pump, radiator, and fan.

OMNITRACTOR.

ENGINE.—Two cylinders, 6$\frac{1}{2}$ × 9in. (165 × 229 mm.), 30 h.p. at 750 r.p.m.
POWER.—20 h.p. (4,000 lb., assuming 200 lb. pull per h.p.).
FUEL.—Paraffin. WEIGHT.—66 cwt.
GEARS.—Two forward (2$\frac{1}{4}$ and 4$\frac{1}{2}$ miles per hour), one reverse. Final drive to wheels by steel pinions.
CAPACITY.—Four furrows, 9-12in. wide and 6-9in. deep, according to land.
WHEELS.—Four. Driving wheels, 5ft. diameter × 1ft. 4in. face ; front wheels, 3ft. 6in. diameter × 9in. face.
BELT PULLEY.—On main engine shaft under control of clutch off side.
GOVERNOR.—Centrifugal.
COOLING.—Pump, radiator, fan, water tank.
OVERALL DIMENSIONS.—Length 11ft. width 7ft.

FARMERS! YOU ARE SAFE

IN USING

SEED CORN, GRASS SEEDS, ROOT SEEDS, MANURES, &c.

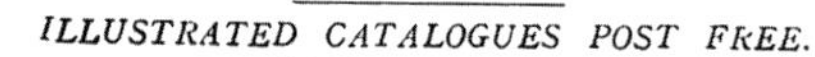

ILLUSTRATED CATALOGUES POST FREE.

WEBB & SONS (Stourbridge) LTD.,
The King's Seedsmen, **STOURBRIDGE.**

Tandem Wire Strainer

(PATENTED).

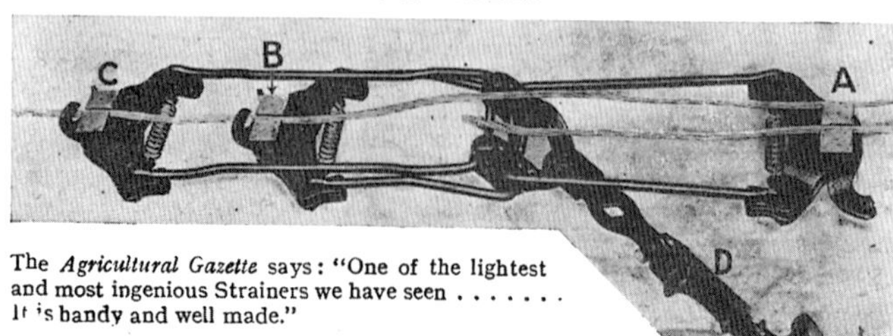

The *Agricultural Gazette* says: "One of the lightest and most ingenious Strainers we have seen .... It is handy and well made."

The TANDEM STRAINER will erect or repair barbed or plain wire fences. It is a one-man outfit—weighs only a few lbs.—and will exert a straining force of 600 lbs. without difficulty owing to the powerful leverage it gives. Does not cut or damage the wire, and can be used for connecting wire to wood or iron posts without drilling or boring holes. Will save its costs in a very short time, and particularly handy for repairing breakages.

Price 30/-

If your Ironmonger or Implement Agent cannot supply you, please write us direct and we will send you full particulars.

ALFRED BULLOWS & SONS, Ltd., Long St. Works, **WALSALL.**

LIST OF PRINCIPAL TRACTOR MAKERS AND AGENTS IN THE UNITED KINGDOM.

Alldays - - -	- Alldays & Onions, Ltd., Matchless Works Birmingham.
Avery - - -	- R. A. Lister & Co., Ltd., Dursley (Glos.).
Bates Steel Mule -	- Austin Motor Car Co., Ltd., Northfield, Birmingham.
B.A.I.C.O. (attachment) - -	- British-American Import Co., Ltd., 11, Haymarket, London, S.W.1.
Beeman - - -	- R. Martens & Co., Ltd., 15a, Wilton St., Grosvenor Place, London, S.W.1.
Boon - - -	- Boon Tractors, Ltd., Eagle Works, Warwick.
Bullock Creeping Tractor	- American Trading Co., 90-93, Fenchurch Street, London, E.C.3.
Burford-Cleveland -	- H. G. Burford & Co., 16, Regent Street, London, S.W.
Caterpillar - -	- Caterpillar Tractors, Ltd., 60, Queen Victoria Street, London, E.C.4.
Clydesdale - -	- R. Martens & Co., Ltd., 15a, Wilton Street, Grosvenor Place, London, S.W.1.
Crawley Agrimotor -	- Crawley Agrimotor Co., Ltd., Mandeville Road, Saffron Walden.
Eagle - - -	- British American Import Co., 11, Haymarket, London, S.W.1.
Emerson - - -	- Brainsbys, Ltd., Broadway, Peterborough.
Eros (attachment) -	- Morris Russell & Co., Ltd., Great Portland Street, London, W.
Farmer Boy - -	- Morris Russell & Co., Ltd., Great Portland Street, London, W.
Fordson - - -	- Henry Ford and Son, Ltd., 59, Shaftesbury Avenue, London, W.
Fowler - - -	- John Fowler & Co., Ltd., Leeds.
Garner - - -	- Henry Garner, Ltd., Moseley, Birmingham.
G.W.W. - - -	- Gaston Williams & Wigmore, Ltd., 212, Great Portland Street, London, W.1.
Heider - - -	- Willys Overland, Ltd., 151, Great Portland Street, W.
Interstate - -	- Austin Motor Car Co., Ltd., Northfield Works, Birmingham.
Ivel, Ivel-Hart - -	- Ivel Agricultural Motors, Ltd., Biggleswade (Beds.).
Kardell - - -	- Power Farm Supply Co., Ellys Road, Coventry.
Killen-Strait - -	- Austin Motor Car Co., Ltd., Northfield Works, Birmingham.
Kingsway - -	- Gaston Williams & Wigmore, Ltd., 212, Great Portland Street, London, W.1.

LIST OF TRACTOR MAKERS (*Continued*).

Lauson, Lion, Little Giant – – –	Marlboro' Motors, Ltd., 404, Euston Road. London, N.W.1.
Lightfoot – – –	Power Farm Supply Co., Priory Lawn Chambers, Ellys Road, Coventry.
Mann – – – –	Mann's Patent Steam Cart & Wagon Co., Hunslet, Leeds.
Marshall – – – –	Marshall, Sons & Co., Ltd., Gainsborough.
Martin Cultivator – –	Martin's Cultivator Co., Ltd., Lincolnshire Ironworks, Stamford.
Mogul – – – –	International Harvester Co. (of Great Britain), Ltd., 80, Finsbury Pavement, London, E.C.
Moline – – – –	British Empire Motors, Ltd., 115, Fulham Road, London, S.W.
Monarch-Neverslip – –	Power Farm Supply Co., Ltd., Priory Lawn Chambers, Ellys Road, Coventry.
Morse – – – –	Fairbanks, Morse & Co., Ltd., 87, Southwark Street, London, S.E.1.
Multipede – – –	Clayton and Shuttleworth, Ltd., Lincoln.
New Simplex – – –	Weeks & Son, Ltd., Waterside, Maidstone.
Omnitractor – – –	Omnitractor Syndicate, Ltd., 18 and 19, Great St. Helen's, London, E.C.3.
Overtime – – –	Overtime Farm Tractor Co., 124-127, Minories, London, E.C.
Petter-Maskell – –	Petters, Ltd., Westland Plough Works, Yeovil.
Saunderson – – –	The Saunderson Tractor & Implement Co., Elstow, Bedford.
Simplex – – – –	Marlboro' Motors, Ltd., 404, Euston Road, London, N.W.1.
Suffolk Punch – – –	Richard Garrett and Sons, Ltd., Leiston, Suffolk.
Summerscales – – –	Summerscales, Ltd., Phœnix Foundry, Keighley, Yorks.
Titan – – – –	International Harvester Co. (of Great Britain), Ltd., 80, Finsbury Pavement, London, E.C.
Victoria – – – –	Walsh & Clark, Ltd., Victoria Works, Guiseley, Leeds.
Wallis – – – –	Ancona Motor Co., Ltd., 78-82, Brompton Road, London, S.W.3.
Whiting-Bull – – –	Whiting (1915) Ltd., 334, Euston Road, London, N.W.1.
Worthington-Ingeco – –	Worthington Pump Co., Ltd., Queen's House, Kingsway, London, W.C.2.
Wyles – – – –	Wyles Motor Ploughs, Ltd., Dursley, Glos.

The above does not purport to be a full list, as the output of some of the leading British engineering firms is necessarily hindered for the present by the war.

INDEX.

ALUMINIUM CASTINGS

FOR ALL AND EVERY TRADE.

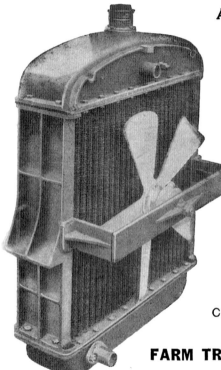

FARM "FILTRATE"

THE OIL
FOR
FARM TRACTORS.

Labour-saving

FARM MACHINERY

We invite enquiries for our British-made

TRACTOR PLOUGHS

(Self-lifting, self-steering, adjustable furrows, one-man outfits),

For the "FORDSON" and other Tractors.

Also ask for particulars of

The "Eros" Combined Land Press and Trailer.
The "Eros" Safeguard for the "Fordson."
Tip Carts, Trailers, Parts for the "Fordson," etc.

THE "EROS" TRACTOR UNIT

converts a "Ford" Car into a powerful Farm Tractor capable of pulling a 2-furrow plough, self-binder, harrow, or other farm implement. Can also be used for road hauling, belt work, etc. Converted back into a "Ford" Car in twenty minutes. Hundreds in use in this country.

Morris, Russell & Co. Ltd.

Agricultural Dept.: 163-5, GT. PORTLAND STREET, W.1.
Head Office: 75, CURTAIN ROAD, LONDON, E.C.2.

Telephones—London Wall 5443-5444. Telegrams—"Luminously, Phone, London."
Works—LONDON, NOTTINGHAM, and SEVEN KINGS.

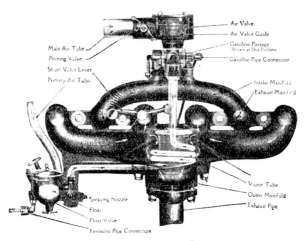

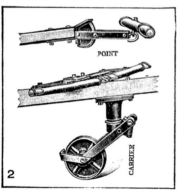

INDEX TO ADVERTISERS.

First published 1918, Iliffe & Sons
First facsimile edition 2012, Old Pond Publishing

ISBN 978-1-908397-20-1

A catalogue record for this book is available from the British Library

Old Pond Publishing Ltd
Dencora Business Centre
36 White House Road
Ipswich IP1 5LT
United Kingdom

www.oldpond.com

Printed in China